APPLICATION OF THERMO-FLUIDIC MEASUREMENT TECHNIQUES

APPLICATION OF THERMO-FLUIDIC MEASUREMENT TECHNIQUES

An Introduction

Edited by

TONGBEUM KIM

School of Mechanical and Aeronautical Engineering
University of the Witwatersrand
Johannesburg, South Africa

TIAN JIAN LU

School of Aerospace Engineering
Xi'an Jiaotong University
Xi'an, China

SEUNG JIN SONG

School of Mechanical and Aerospace Engineering
Seoul National University, Seoul, South Korea

ELSEVIER

AMSTERDAM • BOSTON • HEIDELBERG • LONDON
NEW YORK • OXFORD • PARIS • SAN DIEGO
SAN FRANCISCO • SINGAPORE • SYDNEY • TOKYO
Butterworth-Heinemann is an imprint of Elsevier

BH

Butterworth-Heinemann is an imprint of Elsevier
The Boulevard, Langford Lane, Kidlington, Oxford OX5 1GB, United Kingdom
50 Hampshire Street, 5th Floor, Cambridge, MA 02139, United States

Notices
Knowledge and best practice in this field are constantly changing. As new research and experience broaden our understanding, changes in research methods, professional practices, or medical treatment may become necessary.

Practitioners and researchers must always rely on their own experience and knowledge in evaluating and using any information, methods, compounds, or experiments described herein. In using such information or methods they should be mindful of their own safety and the safety of others, including parties for whom they have a professional responsibility.

To the fullest extent of the law, neither the Publisher nor the authors, contributors, or editors, assume any liability for any injury and/or damage to persons or property as a matter of products liability, negligence or otherwise, or from any use or operation of any methods, products, instructions, or ideas contained in the material herein.

Library of Congress Cataloging-in-Publication Data
A catalog record for this book is available from the Library of Congress

British Library Cataloguing-in-Publication Data
A catalogue record for this book is available from the British Library

ISBN: 978-0-12-809731-1

Publisher: Jonathan Simpson
Acquisition Editor: Simon Tian
Editorial Project Manager: Vivi Li
Production Project Manager: Nicky Carter
Designer: Victoria Pearson

Typeset by Thomson Digital

CONTENTS

CONTRIBUTORS

M.D. Atkins
Atkins Technology, Johannesburg, South Africa

D. Barratt
Department of Mechanical and Process Engineering, ETH Zürich, Zürich, Switzerland

M.F. Boer
School of Mechanical and Aeronautical Engineering, University of the Witwatersrand, Johannesburg, South Africa

S.S. Feng
School of Aerospace Engineering, Xi'an Jiaotong University, Xi'an, China

K. Kim
School of Electrical and Information Engineering, University of the Witwatersrand, Johannesburg, South Africa

T. Kim
School of Mechanical and Aeronautical Engineering, University of the Witwatersrand, Johannesburg, South Africa

T.J. Lu
School of Aerospace Engineering, Xi'an Jiaotong University, Xi'an, China

M.A. Nguepnang
Digital Appliance, Africa Regional Headquarters, Samsung Electronics, Johannesburg, South Africa

D.J. Rossouw
Nuclear Science Division, School of Mechanical Engineering, University of the Witwatersrand, Johannesburg, South Africa

S.J. Song
School of Mechanical and Aerospace Engineering, Seoul National University, Seoul, South Korea

ABOUT THE AUTHORS

Michael D. Atkins is a director of Atkins Tech (PTY) Ltd., Johannesburg, South Africa. He received an MSc (with Distinction) from the School of Mechanical and Aeronautical Engineering, University of the Witwatersrand, Johannesburg, South Africa and has been an associate member of the Royal Aeronautical Society (UK) since 2015. He specializes in fundamental and applied aerodynamics, flow control, thermo-fluidics in gas turbine engines, particle image velocimetry, and planar laser visualization.

Dylan Barratt is a Scientific Assistant and PhD candidate in the Department of Mechanical and Process Engineering (D-MAVT) at Swiss Federal Institute of Technology, Zürich, Switzerland. He received an MSc from Seoul National University, South Korea and is interested in sustainable technologies including particulate capture, gas turbine efficiency, and wind turbine design.

Michael F. Boer is a university lecturer and received an MSc from the School of Mechanical and Aeronautical Engineering, University of the Witwatersrand, Johannesburg, South Africa. He specializes in aircraft design, including low-speed wind tunnel testing with flow visualization.

Shangsheng Feng is a university lecturer and received a PhD from Power Engineering and Engineering Thermophysics, Xi'an Jiaotong University, Xi'an, China. He specializes in fluid flow and heat transfer in pin fin heat sink and porous materials, including temperature and flow field measurement.

Kimoon Kim is a research associate in the School of Mechanical Engineering, University of the Witwatersrand, Johannesburg, South Africa. He received his BEng in Biomedical Engineering and BSc in Electrical Engineering from the Faculty of Engineering and the Built Environment, University of the Witwatersrand, Johannesburg, South Africa. He specializes in software engineering and machine learning.

Tongbeum Kim is a Professor in the School of Mechanical and Aeronautical Engineering at the University of the Witwatersrand, Johannesburg, South Africa.

Tian Jian Lu is a Professor and a former Vice President for Research of Xi'an Jiaotong University, China. Prior to this he was a Chair Professor of Materials Engineering at the University of Cambridge, UK.

Maurel A. Nguepnang is a Junior Technical Manager at Samsung Electronics South Africa. He received an MSc (with Distinction) from the School of Mechanical and Aeronautical Engineering, University of the Witwatersrand, Johannesburg, South Africa. His interests include forced convection in home appliances.

Dillon J. Rossouw is a research associate in the School of Mechanical Engineering, University of the Witwatersrand, Johannesburg, South Africa. He received a BSc in Nuclear Science and Engineering from the Faculty of Science, University of the Witwatersrand, Johannesburg, South Africa. He is specializing in thermal hydraulics in pressurized heavy water reactors.

Seung Jin Song is a Professor in the School of Mechanical and Aerospace Engineering, Seoul National University, South Korea.

PREFACE

The principal purpose of this book is to provide readers with a variety of measurement techniques frequently used in aerodynamics and heat transfer experiments. Many books introducing the principles of each technique are available. However, an easy-to-use book combining the basics, particular test setups, calibration procedures and results, and examples with the physical interpretation of the results has yet to be introduced. The authors hope that, with the introduction of this book, readers can learn both the overall and detailed aspects of each experimental technique in aerodynamics and heat transfer and then apply these to their own work.

The first half of the book (Chapter 2–Chapter 5) addresses the essential techniques in aerodynamics measurements whilst the second half (Chapter 6–Chapter 8) deals with heat transfer measurement techniques. Each section consists of basic principles and calibration procedures/calibration results, followed by fundamental but practical examples. Unless otherwise cited, this book (including examples) is based upon the authors' own work.

Tongbeum Kim, Tian Jian Lu,
and Seung Jin Song
June 2016

CHAPTER 1

Experimentation in Aerodynamics and Heat Transfer

T. Kim*, T.J. Lu, S.J. Song†**
*School of Mechanical and Aeronautical Engineering, University of the Witwatersrand, Johannesburg, South Africa
**School of Aerospace Engineering, Xi'an Jiaotong University, Xi'an, China
†School of Mechanical and Aerospace Engineering, Seoul National University, Seoul, South Korea

1.1 INTRODUCTION

Aerodynamics and heat transfer problems have been historically investigated through three different approaches: mathematical, experimental, and computational. The mathematical approach first gained popularity in the 17th, 18th, and 19th centuries through pioneers like Daniel Bernoulli, Leonard Euler, and Hermann von Helmholtz. However, the complexity of mathematical treatment combined with inadequacy at predicting basic factors like drag prompted a rise in the experimental approach. The Wright Brothers, for example, relied extensively on wind-tunnel testing in the design of their original aircraft. This trend continued through the 20th century as engineers made extensive use of experimentation in the analysis of fluid and heat transfer problems. Most recently, the cost and time requirements of experimentation have encouraged a computational approach, called "computational fluid dynamics (CFD)," to fluid and heat transfer problems in which the governing equations are solved numerically. However, the computational expense of solving the governing equations in their original form is impractical for all but the simplest configurations, leading to approximate methods of solution often including turbulence models. Furthermore, experimental data are still needed to validate the CFD results. Therefore, a sound knowledge of experimentation remains a necessary asset for not only *experimentalists* but also those interested in heat and fluid problems. This book provides such knowledge associated with aerodynamics and heat transfer fields. The techniques dealt with in this book are limited to those thought to be essential tools for researchers in these two disciplines.

For any experimental studies, a fundamental question is perhaps "what to measure." Answering this question is associated with *governing equations* that need to be closed. The governing equations in aerodynamics and heat

Application of Thermo-Fluidic Measurement Techniques
http://dx.doi.org/10.1016/B978-0-12-809731-1.00001-0

transfer are the "continuity equation," "momentum equations," and "energy equation." These equations contain terms such as static pressure, velocity, temperature, and their gradients. The measurement techniques in this book attempt to quantify them.

On the other hand, in some cases, it is not possible, at present, to experimentally obtain the parameters required to close the above-mentioned governing equations due to experimental limitations. Furthermore, there are parameters in the governing equations that may not be experimentally quantified without compromising data quality, as could be done mathematically.

The following sections in this chapter aim not to re-derive the governing equations in aerodynamics and heat transfer but to show their physical significance.

1.2 AERODYNAMICS

The motion of fluid particles is governed by *Navier–Stokes* (*N–S*) equations. In addition, the fluid has to satisfy the "continuity equation" stating mass conservation for a given volume of interest (control volume). In one-dimensional (1D) form, along the x-axis, the general expressions of these equations for incompressible and steady flow without body force may be expressed as:

- Continuity equation:

$$\rho \frac{du}{dx} = 0 \tag{1.1}$$

- Momentum equation:

$$\rho u \frac{du}{dx} = -\frac{dp}{dx} + \mu \frac{d^2u}{dx^2} \tag{1.2}$$

where u is the velocity component in the x-direction, p is the static pressure, ρ is the density of fluid, and μ is the viscosity of fluid. Details of the final expression of the equations including mathematical derivations can be found in the references (Long, 1999; Bejan, 2003; Kaviany, 2001; Incropera and DeWitt, 1990).

As seen in Eqs. (1.1) and (1.2), solving continuity and momentum equations *experimentally* is about how to close the terms such as the pressure gradient (dp/dx), velocity (u), and velocity gradient (du/dx) and its gradient (du^2/dx^2). For example, to experimentally obtain the pressure gradient

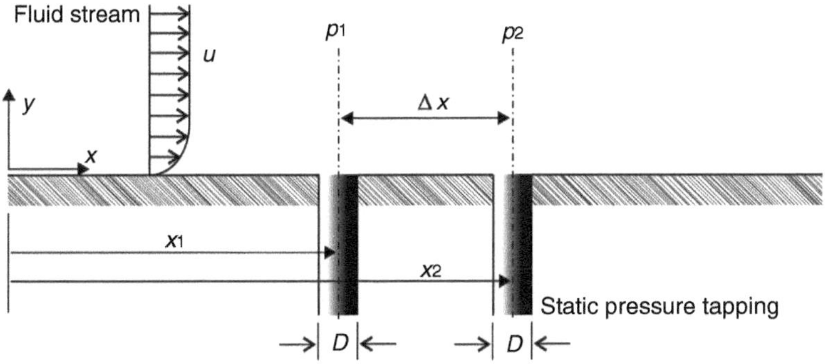

Figure 1.1 ***Schematic of static pressure tappings flush with a solid surface.***

(dp/dx) term in Eq. (1.2), static pressures at two points (p_1 and p_2), with an interval between the two pressure tappings (dx) along the direction of flow, have to be measured (Fig. 1.1). In experiments, a commonly and practically useful form of the definition of the pressure gradient may be expressed as:

$$\frac{\Delta p}{\Delta x} = \frac{p_1 - p_2}{x_2 - x_1} \quad (1.3)$$

In the actual experimental setup, the closest interval (Δx) might be the diameter of the pressure tappings (D), which is typically on the order of millimeters.

For detailed gradient measurements, if many pressure tappings are aligned along the x-axis with "a very fine interval" (eg, ~ the diameter of the pressure tappings), another unexpected problem may arise. The flow past downstream tappings will be affected by those upstream because the upstream tappings may act as cavities and alter the flow. There may exist a minimum interval (between the two pressure tappings placed in an in-line configuration along the x-axis) which minimizes such interference effects without sacrificing data resolution.

Similarly, the velocity gradient terms in Eq. (1.2) can be obtained by measuring velocity at different locations (separated by Δx) using various techniques (eg, Pitot tube, Laser Doppler Anemometry (LDA)). The accuracy of the measurements depends strongly on Δx and its relative scale to the overall flow domain of interest and the probe size. Furthermore, when internal flow in a curved channel (eg, a turbine blade passage illustrated in Fig. 1.2) is of interest, the measurement of the velocity (gradient) is hindered also by the accessibility of some experimental techniques. Some pneumatic probes (eg, Pitot tube) do not have enough traversing

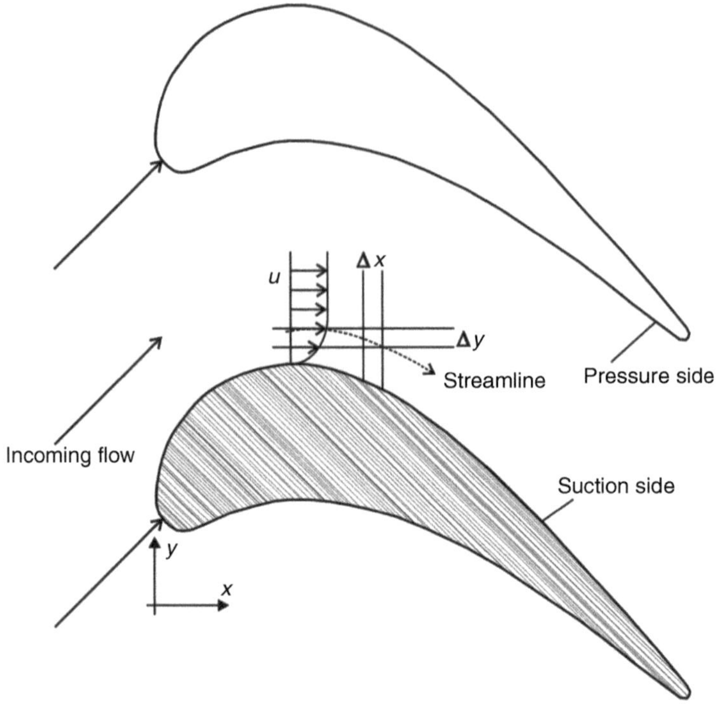

Figure 1.2 ***Internal flow velocity measurement in a turbine blade passage.***

space and optical techniques, for example, LDA, cannot derive results in a region of shadow. Thus, experimental data are strongly dependent upon measurement methods. Therefore, it is important to question whether an appropriate experimental technique that does not contaminate main features of interest and is experimentally accessible to the flow field, is being used.

1.3 CONVECTION AND CONDUCTION HEAT TRANSFER

Heat is transferred by conduction, convection, radiation, or by their combination(s). Conduction occurs in a stationary medium, convection requires a moving fluid, and radiation can occur regardless of medium type. In all three cases, temperature difference is the driving force for the transfer of heat. Heat is transferred from a high temperature point (region) to a low temperature point (or region). In this book, radiation is not covered because, at moderately low temperatures, conduction and convection usually dominate, often occurring together.

1.3.1 Convection Heat Transfer

Convection heat transfer is governed by the motion of fluids. Its temperature field is expressed by the energy equation whose simplified form, assuming incompressible fluid and constant thermal conductivity (k) with negligible viscous dissipation, may be given in a 2D (x–y) plane as:

$$\rho C_p\left(u\frac{\partial T}{\partial x}+v\frac{\partial T}{\partial y}\right)=k\left(\frac{\partial^2 T}{\partial x^2}+\frac{\partial^2 T}{\partial y^2}\right) \tag{1.4}$$

Experimentally solving the energy equation, Eq. (1.4), involves not only temperature measurements for the $\partial T/\partial x$, $\partial T/\partial y$, $\partial^2 T/\partial x^2$, and $\partial^2 T/\partial y^2$ terms but also velocity measurements for the u and v terms.

In many convection heat transfer applications, the convective fluid motion acts to cool down or heat up a solid object into which heat is transferred at a moving fluid and solid interface. The resultant heat transfer at the interface is of most interest. In such a case, it is more convenient to define an expression to quantify heat convection at the interface as:

$$Q = hA\left(T_s - T_f\right) \tag{1.5a}$$

where

$$h=\frac{q''}{\left(T_s - T_f\right)} \tag{1.5b}$$

Here, q'' is the heat flux, h is the convection heat transfer coefficient, which is not a property of fluids but of flow, T_s is the solid surface temperature, and T_f is the moving fluid stream temperature adjacent to the solid surface (to be specified later). Unlike thermal conductivity of a fluid, the heat transfer coefficient, h, depends on many parameters including fluid velocity, states of flow (turbulent or laminar), surface geometry (rounded or flat), and fluid properties (viscosity and density).

To elaborate on heat convection, consider the case illustrated in Fig. 1.3a where, by imposing heat through the wall of an empty channel, the local heat transfer coefficient on the surface of the wall can be calculated (measured) at each stream-wise station, x. At the solid boundary, heat transfer takes place only through conduction since there is no relative motion between the surface and the fluid particles that are in contact with the solid surface. This is the so-called "no-slip condition." Away from the solid

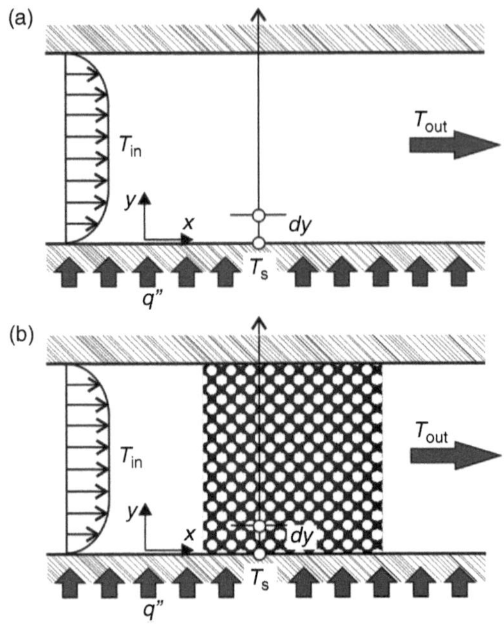

Figure 1.3 ***Schematic description of convection heat transfer.*** (a) In an empty channel; (b) in a porous channel.

surface, the transfer of heat is governed by convection. The local convection heat transfer coefficient $h(x)$ can hence be obtained as:

$$h(x) = k_s \left.\frac{dT}{dy}\right|_{y=0} \quad \text{at an arbitrary } x \tag{1.6a}$$

To experimentally solve Eq. (1.6a), two temperature values have to be measured [ie, $T_s(y = 0)$ and $T_f(y \sim 0)$] at a particular stream-wise location, for example, $x = x_1$, as:

$$h(x_1) = k_s \left.\frac{T_s(x_1) - T_f(x_1)}{dy}\right|_{y=0} \quad \text{at } x = x_1 \tag{1.6b}$$

Here, the solid surface temperature $T_s(x_1)$ can be measured by a thermocouple mounted flush with the surface. On the other hand, measuring T_f at the same x station right above the solid surface ideally at $y = 0$ but practically near the solid surface ($y = dy$), is questionable. The selection of "near the solid surface" requires judgement. The term "near" may be determined by the diameter of bead-type thermocouples, typically on the order of 100 μm. The situation becomes worse if a porous material (as illustrated in Fig. 1.3b) is inserted into the flow channel, which sometimes occurs inside heat exchange devices. The measurement of fluid temperature inside a porous material is presently very challenging.

For practical purposes, different definitions of the convection heat transfer coefficient other than that given in Eq. (1.6a) have been used, basically depending on the reference temperature in Eq. (1.5b). According to Azar and Moffat (1991), typically three reference temperatures have been popular choices: the inlet fluid temperature (T_{in}), the mean fluid temperature (T_m), and the adiabatic temperature (T_{ad}), resulting in:

$$h_{in} = \frac{q''}{\left(T_s - T_{in}\right)} \tag{1.7a}$$

$$h_m = \frac{q''}{\left(T_s - T_m\right)} \tag{1.7b}$$

$$h_{ad} = \frac{q''}{\left(T_s - T_{ad}\right)} \tag{1.7c}$$

where $T_m = T_{in} + Q/(\dot{m}C_p)$, T_{ad} is the adiabatic temperature the solid surface would achieve if there were no heat transfer to (or from) it. The adiabatic condition is theoretically important but difficult to realize in practical situations. Here, $\dot{m}$ is the mass flow rate (kg/s). It has been experimentally observed (Azar and Moffat, 1991) that the following (Eq. (1.8)) always holds for forced convection with one heated wall:

$$h_{ad} > h_m > h_{in} \tag{1.8}$$

As long as the definition of h is clearly stated and the difference among the definitions is acknowledged by the users, any one of Eq. (1.7a–c) can be used without confusion or conflict with other users.

The dimensional convection heat transfer has units of W/(m^2K) and depends on the heat flux and temperature difference across the surface of concern. Nondimensional parameters are often preferable. The commonly used nondimensional groups associated with convection and conduction are the Nusselt number and the Biot number, which are defined as follows:

- Nusselt number:

$$Nu = \frac{h}{k_f/C} \tag{1.9a}$$

- Biot number:

$$Bi = \frac{h}{k_s / C} \tag{1.9b}$$

The Nusselt number (Nu) is the ratio of the convection portion of the total heat transfer to the conduction portion through the fluid. The thermal field in the fluid side is characterized by the Nusselt number. It is a product of the convection heat transfer coefficient (h), thermal conductivity of the fluid (k_f), and a characteristic length scale (C). For example, in the case of heat transfer around a single cylinder, the cylinder diameter is conventionally used as the length scale. If some heat-exchanging materials are inserted in a flow channel, a hydraulic diameter (if the channel is rectangular) or pipe diameter (if the pipe is circular) is used.

The Biot number (Bi) is a measure of the uniformity of temperature in solids. It is a product of the convection heat transfer coefficient (h), thermal conductivity of the solid (k_s), and a characteristic length scale (C). The temperature change at the surface of a solid block will affect the distribution of temperature inside the block. In general, the temperature distribution within the solid block depends on the extent of temperature variation at the block surface, thermal conductivity of the solid block material, and size of the block. A smaller-value Biot number (typically Bi $<$ 1) means that a large portion of the solid becomes thermally "uniform" (Long, 1999).

1.3.2 Conduction Heat Transfer

Conduction is the mode of heat transfer that occurs as a result of the motion of molecules, electrons, and vibrations in the atomic lattice. For solids that are poor conductors, heat is transported through translational vibrations of the lattice structure. In good conductors, the process is greatly enhanced by the motion of free electrons. In a stationary fluid (which is only possible when a temperature difference within the fluid is small enough not to induce a density difference that also can drive the fluid's motion, that is, natural convection), heat is transferred via conduction. The mechanism for conduction in liquids and gases is similar. In liquids and gases, the temperature of a molecule is related to its kinetic energy. Molecules in high-temperature fluids and gases have greater mean velocities and thus higher kinetic energy. The transfer of energy takes place through collisions between higher kinetic energy molecules and those with low kinetic energy (Long, 1999). Furthermore, particle vibration in fluids includes both translational and rotational modes which contrast with the purely translational vibration of solids.

The temperature field in a medium determined by such microscopic events can be obtained also by the energy equation, Eq. (1.4) with no convective motion (eg, a solid or a stagnant fluid), where $u = v = 0$, yielding:

$$\frac{\partial^2 T}{\partial x^2} + \frac{\partial^2 T}{\partial y^2} = 0 \tag{1.10}$$

which is called "Laplace's equation." Details of the final expression of the equation including mathematical derivation can be found in the references (Long, 1999; Bejan, 2003; Kaviany, 2001; Incropera and DeWitt, 1990). Experimentally solving the energy equation for conduction according to Eq. (1.10) involves only temperature measurements for the $\partial^2 T/\partial x^2$ and $\partial^2 T/\partial y^2$ terms.

These microscopic events manifest themselves at the macroscopic scale by Fourier's law: that the heat transfer per unit area, q'', or heat flux in a given direction is proportional to the temperature gradient in that direction. This is written in general terms as:

$$q_n \propto \frac{\partial T}{\partial n} \tag{1.11}$$

where n denotes the direction of the outward normal. To be specific for the x-direction we would write for an area A that

$$\frac{Q_x}{A} = q_x = -k\frac{\partial T}{\partial x} \qquad (\mathrm{W/m^2}) \tag{1.12}$$

where Q_x is the rate of heat transfer in watts. A leading negative sign in Eq. (1.12) ensures that flux is positive in the direction of increasing x and the constant of proportionality k is known as the thermal conductivity.

Fig. 1.4a schematically illustrates the conduction of heat through a solid block of thickness L. Fourier's law (Eq. 1.12) states that heat is transferred from one temperature T_0 to the other temperature T_L if $T_0 > T_L$ holds. With an increasing amount of heat flux (q or q'') entering the solid block, the temperature difference ($\Delta T = T_0 - T_L$) becomes larger and is proportional to the imposed heat flux. This proportionality is called "thermal conductivity (k)," typically assumed to be constant with temperature. However, as shown in Fig. 1.4b, this is not always true since thermal conductivity, which is a property of solids (or fluids), varies with temperature. Only in a certain range of temperature (eg, that marked as the "valid range" in Fig. 1.4b), is

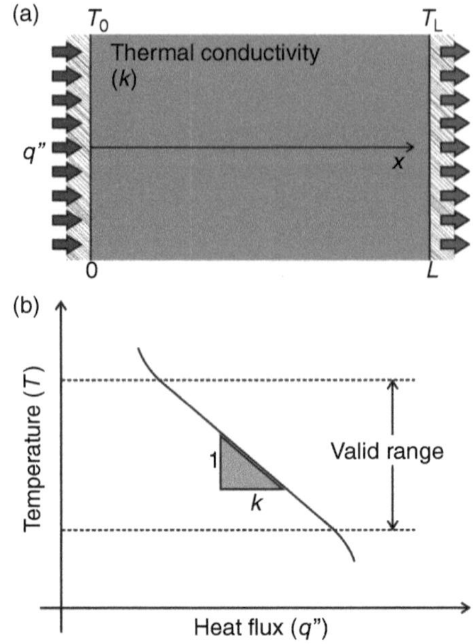

Figure 1.4 ***Conduction heat transfer.*** (a) Heat conduction through a solid plate; (b) variation of thermal conductivity with nominal operating temperature.

the thermal conductivity k constant, making it independent of temperature in Eq. (1.12).

Fig. 1.5 shows the variation of thermal conductivity of, for example, aluminum (Al) with the absolute temperature from 1 to 1000 K. It can be

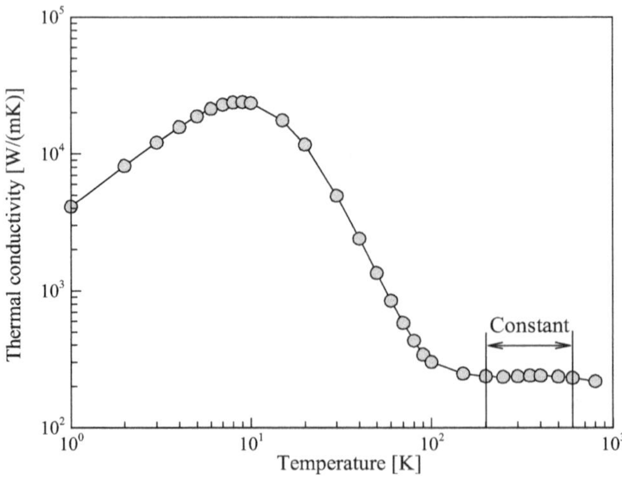

Figure 1.5 ***Variation of thermal conductivity of aluminum (Al) with absolute temperature from 1 K to 1000 K.*** *Source: (Data from http://www.efunda.com/materials/elements/TC_Table.cfm?Element_ID=Al).*

noted that the thermal conductivity is largely constant from 200 to 600 K, which roughly coincides with typical engineering operating temperatures. Each material has a different dependence of thermal conductivity on temperature. Such dependence may lead to a significant error being unexpectedly produced during the data process. Therefore, it is advised to be aware of the temperature dependence of a material's conductivity before conducting any heat transfer measurement.

1.4 CLASSIFICATION OF MEASUREMENT TECHNIQUES

A variety of measurement techniques may be classified according to the nature of contact between the medium of interest and the measurement devices (Childs, 2001) as:

1. Invasive: the transducer (or probe) is in direct contact with the medium of interest, for example. thermocouples, pressure tapping (or probe),
2. Semiinvasive: the surface of interest is treated in a certain way to enable observation at a remove, for example, thermochromic liquid crystal (TLC), infrared (IR) thermometry (if surface treatment is made to modify emissivity),
3. Noninvasive: the medium of interest is observed remotely, for example, laser Doppler anemometry (LDA), particle image velocimetry (PIV) (if seeded particles do not alter the flow field), and IR thermometry.

The advantage in invasive techniques is associated typically with their robustness and reliability of data accuracy. However, due to the physical presence of the probes, the flow and temperature fields may be affected. On the other hand, it is reasonable to assume that the effects are negligibly small if the field of interest is large relative to the size of the probes. Generally, the invasive techniques are used for point measurements. To obtain field information, the invasive probes should be traversed, which is rather time consuming.

Semi-invasive techniques are usually used for surface field measurements. For example, pressure-sensitive paints and thermochromic liquid crystals (TLCs) are often painted on the surface of interest, and are composed of special particles that react to pressure or temperature. The size of TLC particles is on the order of 1/10th of a μm (Ireland et al., 1995). Therefore, if the area of interest is comparable to that of the TLC particles, the flow field is greatly affected.

Noninvasive techniques appear to be ideal since there is no disturbance of the flow and temperature fields by the physical bodies of probes during experimental measurements. Typically, optical measurements such as laser sources are involved. When complex geometries are concerned, however, optical access can be problematic, in addition to being more costly than the invasive techniques. Furthermore, optical techniques are vulnerable to reflection which prohibits measurements near a solid surface, for example, inside the boundary layer. Therefore, which technique is selected in practice depends on a variety of issues such as availability, cost, time constraint, and test configuration.

1.5 CLOSURE

This book introduces widely and commonly used measurement techniques in aerodynamics and heat transfer. The accuracy of the obtained measurement data is dependent upon the proficiency of users with a given technique. In each chapter, we give brief but helpful comments for each technique based on our own experience. Furthermore, working examples on selected fundamental topics are provided.

REFERENCES

Azar, K., Moffat, R.J., 1991. Heat transfer coefficient and its estimation in electronic enclosures. Proceedings of the National Electronic Packaging and Production Conference, Boston, MA, USA, pp. 361–372.

Bejan, A., 2003. Convection Heat Transfer, third ed. Wiley, New York.

Childs, P.R.N., 2001. Practical Temperature Measurement. Butterworth-Heinemann, Oxford.

Incropera, F.P., DeWitt, D.P., 1990. Fundamentals of Heat and Mass Transfer, third ed. Wiley, New York.

Ireland, P.T., Wang, Z., Jones, T.V., 1995. Measurement Techniques: Liquid Crystal Heat Transfer Measurements, von Karman Institute for Fluid Dynamics, Lecture Series 1995-01.

Kaviany, M., 2001. Principles of Convection Heat Transfer, second ed. Springer, Berlin.

Long, C.A., 1999. Essential Heat Transfer. Pearson Education Limited, Harlow, UK.

PART 1

Measurements in Aerodynamics

CHAPTER 2

Flow Visualization

M.D. Atkins*, M.F. Boer**
*Atkins Technology, Johannesburg, South Africa
**School of Mechanical and Aeronautical Engineering, University of the Witwatersrand, Johannesburg, South Africa

2.1 INTRODUCTION

The visualization of flow patterns provides qualitative and graphical details of flow processes rather than quantitative flow data. Firstly, a flow field can be indirectly realized by visualizing the footprint of the flow over a surface, using surface flow visualization techniques. Secondly, a flow field can be directly visualized, using in-flow visualization techniques. The purpose of this chapter is to demonstrate methods of visualizing a flow field from both sets of techniques.

Six different methods are considered, including:

- Surface flow visualization using the oil-dye technique
- Surface flow visualization using thermochromic liquid crystal
- Surface flow visualization using infrared thermography
- In-flow visualization using neutrally buoyant helium bubbles
- In-flow visualization using planar laser imaging
- In-flow visualization using ink-dye pigment injection

Other flow visualization techniques can be found in a book edited by Smits and Lim (2000).

It is worth noting that flow images do not provide any meaningful information on their own unless the investigator interprets the images. This process is called the "physical data interpretation." Furthermore, the obtained images may look different depending on the user's individual approach and expectation. The descriptions of the actual implementation of each technique that follow include problem description, test parameters, and interpretation of the obtained data.

2.2 SURFACE FLOW VISUALIZATION USING THE OIL-DYE TECHNIQUE

2.2.1 Background

The characterization of fluid motion using the oil-dye technique is one of the oldest visualization techniques in aerodynamics. This technique is

Application of Thermo-Fluidic Measurement Techniques
http://dx.doi.org/10.1016/B978-0-12-809731-1.00002-2

cheap to implement but can only provide qualitative information. The basic principle associated with this flow visualization technique is that the aerodynamic shear stresses redistribute the mixture of oil and dye pigment coated on a surface of interest, causing the mixture to move in the direction of the wall shear stress vector, and in doing so, to create a flow pattern on the surface. What the mixture shows as a result of the flow pattern above the coated surface is the image of streaklines.

2.2.2 Dye Particles and Oil

The implementation of the oil-dye technique begins with the selection of a dye pigment and an oil to mix. An easily purchasable dye pigment may contain fluorescent dye particles having a nominal diameter of 1 μm. Such dye particles have been used in the field of forensic science when identifying fingerprints left on objects.

In low-speed applications, light diesel oil is preferred due to its relatively low viscosity. When light diesel oil and fluorescent pigment are selected, the mixing ratio of oil to dye pigment is strongly dependent upon flow velocity. In the mixture, too much dye relative to the oil tends to make detailed flow features such as the location of the "saddle point" unclear because the dye particles are aggregated. On the other hand, an insufficient amount of dye particles leaves no trace of endwall flow features. Under a given flow velocity, the actual value of the mixing ratio is best selected via trial and error.

2.2.3 Gravity Effect

For a flat (endwall) surface which is horizontally configured, the dye particles on the surface are subjected to the same gravitational force. Therefore, the resultant flow patterns can be thought to be mainly caused by flow above the flat surface. However, when the endwall surface of interest is curved or vertically oriented, special care must be taken to ensure that no gravity effect changes the flow patterns on it.

At high Reynolds numbers (or to be exact, with sufficiently high flow velocity), it may be thought that the fluid momentum can overcome the gravitational force. However, the flow needs to be stopped during photographing. Then, the formed streaklines may be altered in the direction of the gravitational axis.

2.2.4 Surface Preparation and Photographing

Before the application of the mixture, the surface is painted typically black. This process is particularly useful when the surfaces are metallic. The main purpose of painting is to enhance the reflectivity when an ultraviolet light having a typical wavelength of 366 nm is used to illuminate the surface flow pattern while it is being photographed. For photographing, the visualized surface is removed from the experimental setup. In addition, manually adjusting the focal length of a camera would give better images.

Example 2.1: Endwall Flow Around an Inclined Circular Cylinder

A. Background

Flow around a circular cylinder which is mounted normal to a flat plate is a classical problem in fluid mechanics and heat transfer. When, however, the cylinder is inclined with respect to the direction of flow, flow properties around the cylinder and near the endwall are changed. An easy indication can be found from the endwall surface flow pattern. This example deals with "how the inclination of a single cylinder with respect to the direction of the mainstream affects endwall surface flow features." Details of the physical understanding of this problem as well as the experimental procedures associated with the oil-dye technique are discussed.

B. Test Setup and Cylinder Configurations

Single circular cylinders with selected inclination angles of $\alpha = 0, 15, 30$, and 45° have been tested. The inclination angle (α) is measured from the z-axis (Fig. 2.1), which is normal to the direction of the mainstream. A positive inclination angle indicates counterclockwise rotation with respect to the y-axis (along the channel width). Thus, the cylinder is rotated clockwise as its inclination angle increases.

Typically, two different methods can be used to change the inclination of a cylinder. One is to fix both ends of the cylinder to the endwall plates.

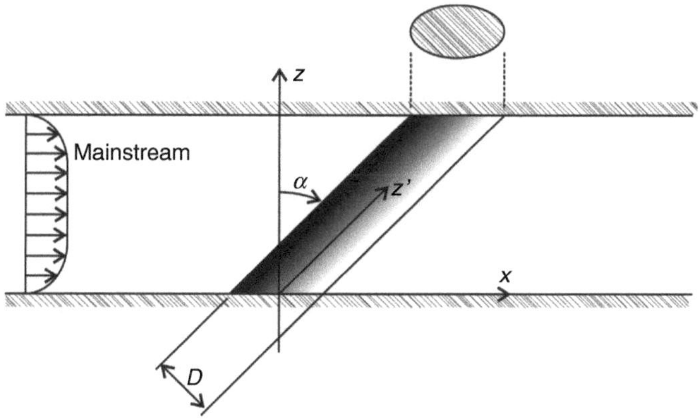

Figure 2.1 ***Configuration of a single cylinder inclined with respect to the direction of fluid flow.***

Table 2.1 Parameters of Cylinders Tested for Flow Visualization in a Wind Tunnel

Inclination angle α	Cylinder			Channel		Blockage ratio D/W
	Diameter, D (mm)	Length, l (mm)	l/D	Width, W (mm)	Height, H (mm)	
0°	30	80.0	2.67	200	80	0.15
15°		82.8	2.76			
30°		92.4	3.08			
45°		113.1	3.77			

When the lower endwall plate is shifted longitudinally and upward, the angle between the mainstream and the cylinder axis increases. During this inclination, the cylinder length remains unchanged. Another way, which is employed in this example, is to fix the channel height (H) and rotate the cylinder with respect to the transverse axis. The consequence is that the length (l) of the cylinder sandwiched between the channel endwalls is increased according to $l = H/\cos(\alpha)$. The parameters of the cylinder models and the test channel are listed in Table 2.1.

C. Endwall Surface Flow Patterns

The topology of the endwall surface flow pattern and horseshoe vortices around a cylinder in an upright position in a turbulent flow regime has widely been studied (Baker, 1979, 1980; Ballio et al., 1998; Hung et al., 1991; Hunt et al., 1978; Khan et al., 1995; Khan and Ahmed, 2005; Simpson, 2001; Tobak and Peake, 1979; Visbal, 1991). This section aims to review the dominant features together with the data obtained from the present example as a reference.

The mean characteristics of surface flows around a single cylinder placed normal to the endwall in Fig. 2.2a demonstrate a well-known horseshoe vortex system clearly marked by oil-dye mixtures at $Re_D = 7.8 \times 10^3$. Since only the time-averaged behavior of the horseshoe vortex system is captured, the flow pattern (horseshoe vortex) is symmetric with respect to the cylinder axis.

For physical interpretation of the flow images, topological models have been proposed (Hunt et al., 1978; Tobak and Peake, 1979). A common feature suggested is the saddle point (S), which refers to a point through which two particular streamlines pass. Each line acts as a barrier in the field of limiting streamlines so that one set of streamlines cannot cross another set of streamlines. A saddle point which is the position of the primary separation (λ_s) is located approximately 1.5D upstream from the cylinder axis (or 1D from the leading edge). This value at $Re_D = 7.8 \times 10^3$ lies on the higher side of the data range ($0.8D < \lambda_s < 1.5D$) (Visbal, 1991). In this photograph, the separation lines are indicated by brighter lines as a result of the accumulated dye particles. On the upper half of the flow image, the attachment (dotted) and separation (solid) lines are sketched over the image.

In summary, the topology of the vortex patterns and the corresponding limiting streamline patterns are sketched together in Fig. 2.2b. The attachment and separation lines are inferred from the endwall flow image of Fig. 2.2a. The appearance of the vortex pattern with the three-vortex system is based on the measured flow pattern and adapted from Baker (1979) and Visbal (1991). Here, for the endwall flow patterns, the saddle point (*S*) is drawn as the line of separation in turbulent flow (Baker, 1980). The foci (*F*) marked downstream of the cylinder represents the spiraling of stream-

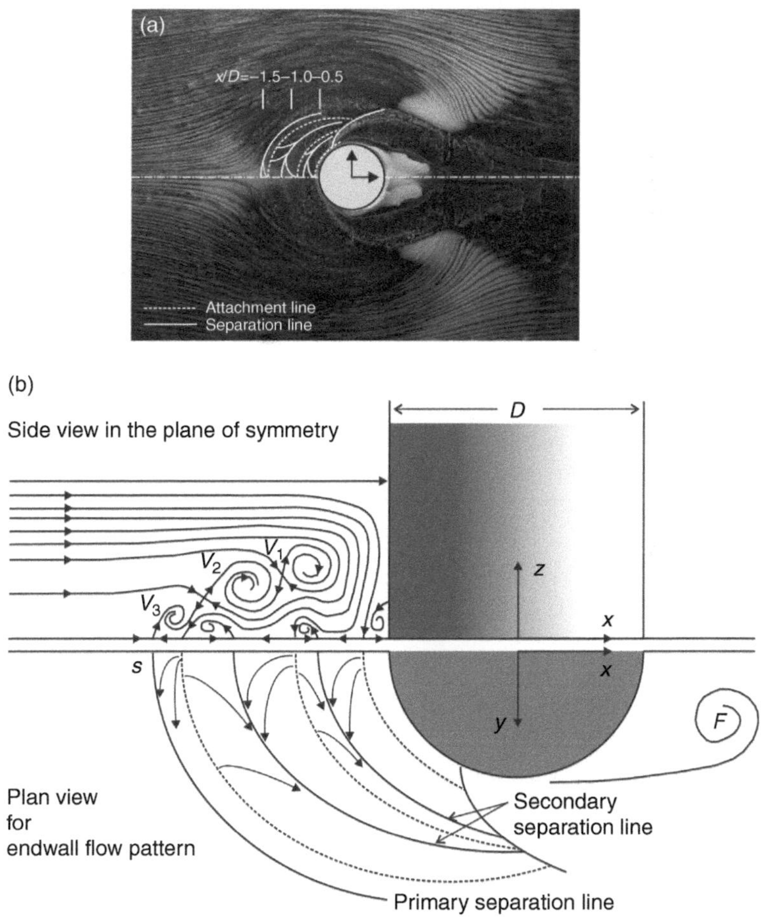

Figure 2.2 ***Flow around an upright circular cylinder ($\alpha = 0°$).*** (a) Visualized endwall flow (Choi et al., 2007); (b) a composite sketch of vortex patterns in a three-vortex system and the corresponding limiting streamline patterns (topological description adapted from Baker (1979) and Visbal (1991). with the saddle point representing a line of separation).

lines into a central point which occurs due to flow separation from the cylinder. The outermost separation line denoted as "*S*" is generated by the flow separating from the endwall. The counter-rotating vortex near V_3 is responsible for the second separation line while the one beneath V_1 causes the third separation line.

Following the configuration shown in Fig. 2.1, the inclination angle is varied from 0 to 45° with a 15° increment at $Re_D = 7.8 \times 10^3$. The mean flow features are shown in Fig. 2.3a–c. In comparison with the upright case, distinct changes are visible in the overall shape of the horseshoe vortices around the cylinder. The appearance of the more streamlined pattern of horseshoe vortices is attributed to the change in the cross-sectional shape of the cylinder from circular to elliptical. For example, at $\alpha = 15°$, the saddle point (*S*) moves closer to the leading edge, for example, located approximately at $\lambda_s = 1.4D$ in comparison with that at $\lambda_s = 1.5D$ for the upright case. A further increase in the inclination angle to 30 and 45° causes the saddle point to move even nearer the leading edge, for example, $\lambda_s = 1.25D$ for $\alpha = 30°$ and $\lambda_s = 1.2D$ for $\alpha = 45°$. Since the distance from the cylinder axis to the cylinder edge changes according to the inclination of the cylinder, the actual distances (L_s) from the primary separation point to the cylinder edge can be estimated from the flow images, as:

$$L_s = \lambda_s\big|_{measured} - \frac{1}{2\cos(\alpha)} \tag{2.1}$$

Here, L_s is calculated to be 0.88, 0.67, and 0.49*D*, respectively, for $\alpha = 15$, 30, and 45°; in comparison, $L_s = 1.0D$ for the upright case. This clearly indicates the significant thinning of the horseshoe vortex system in the plane of symmetry as the inclination angle is increased.

2.3 SURFACE FLOW VISUALIZATION USING THERMOCHROMIC LIQUID CRYSTAL

2.3.1 Background

Thermochromic liquid crystal (TLC) is a substance that exhibits a definite color at a specific temperature. This feature has been explored by many researchers to study the heat transfer characteristics of surfaces. The thermochromic liquid crystal takes advantage of the analogy between heat transfer and wall shear; the gradients in wall shear stress mimic those in heat transfer rates. The visualized image of such surface flow patterns is, in fact, a by-product which represents an instantaneous snapshot of a series of images captured during transient thermal mapping with TLC. Since the details regarding TLC are discussed in Chapter 7, only a brief summary covering the essential characteristics of TLC is given here.

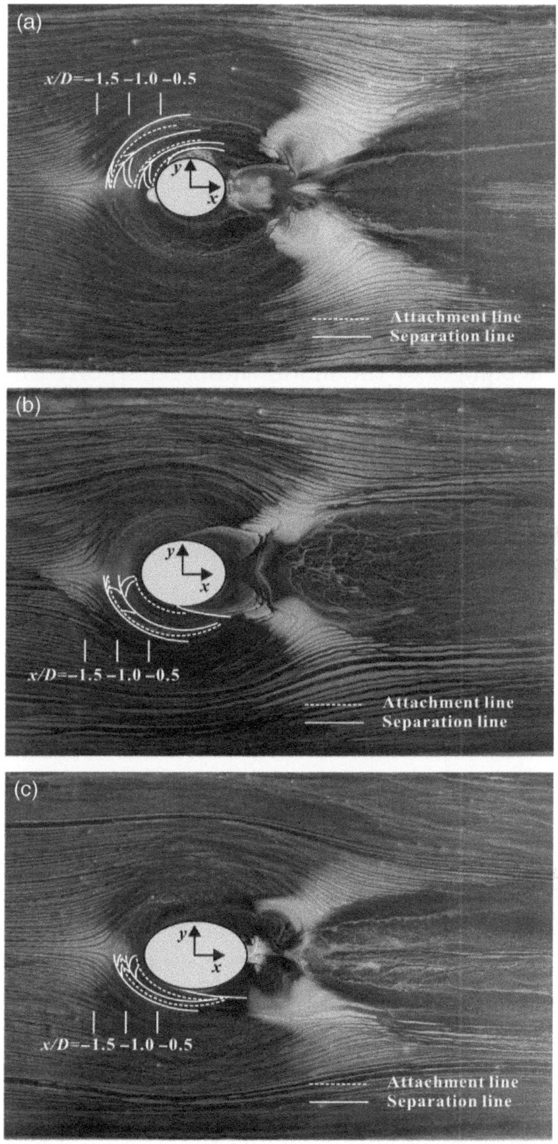

Figure 2.3 ***Endwall surface flow images around a single cylinder having backward inclination, with streaklines superimposed on the flow patterns visualized by the oil-dye mixture technique at $Re_D = 1.0 \times 10^4$ (Choi et al., 2007).*** (a) $\alpha = 15°$; (b) $\alpha = 30°$; (c) $\alpha = 45°$.

2.3.2 Application

The solution mixture of TLC, premixed binder, and deionized water is sprayed using an airbrush. The pressure of the filtered compressed air is best selected by trial and error for a given distance between the airbrush nozzle and the target surface. The typical thickness of a coated TLC layer is approximately 20 μm (Baughn et al., 1999).

2.3.3 Illumination

At a given temperature, the reflected light may vary with the type, intensity, and incidence angle of the illumination. As reported in Camci et al. (1992), the dependence on the incident angle will not be significant if it is arranged to be within ±35° with respect to an axis normal to the surface. Moreover, the source of illumination must not radiate additional heat to the surface, otherwise the radiated heat from the lighting will change the surface temperature pattern that is supposed to mimic the shear stress pattern (flow pattern).

2.3.4 Heating

Two different heating methods (ie, steady-state and transient) can be used, as illustrated in Fig. 2.4, to induce color reflection of the TLC particles. For steady-state heating (Fig. 2.4a), a uniform and constant heat flux is imposed from (eg, gold) film attached to a solid plate (typically the endwall with low thermal conductivity). In this case, heat is transferred from the endwall to the fluid. A black backing paint is sprayed on the heating film to enhance its reflectivity, and a TLC layer is coated on top of the paint using an airbrush. The flow pattern above the TLC layer changes the temperature distribution on the endwall, the image of which is captured with a camera.

For transient heating (Fig. 2.4b), a sudden rise of flow temperature generates a thermal pulse traveling through the width of the endwall. This thermal boundary condition allows the heat transfer coefficient on the endwall surface (assumed to be semiinfinite) to be determined. To realize this sudden rise of fluid temperature, a heating device is typically employed. A transparent (acrylic) endwall plate is typically used. Again, TLC is sprayed on the inner surface of the endwall, on top of the black paint. A camera views the color change of the TLC coating through the transparent endwall (see the following example and Chapter 7 for details).

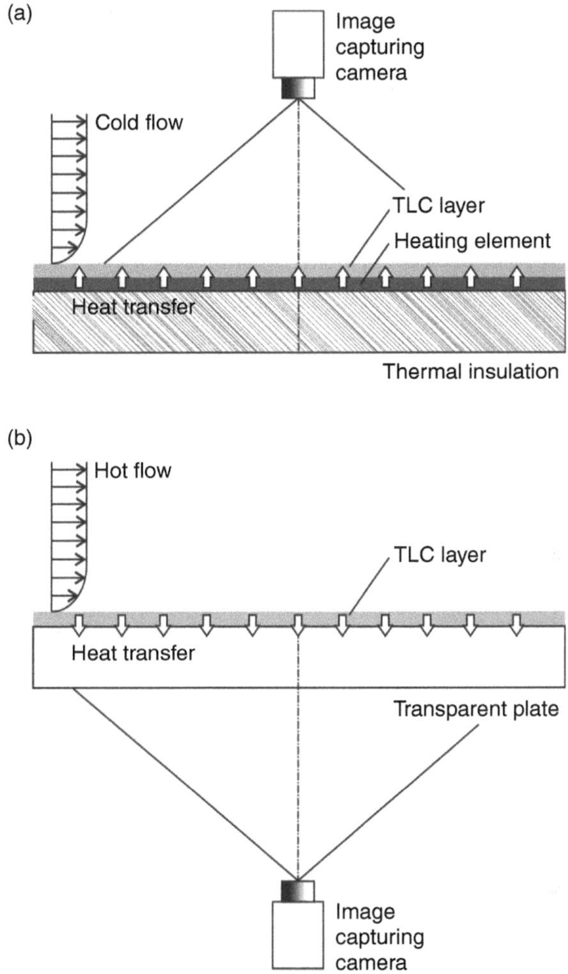

Figure 2.4 ***Heating methods to cause color reflection of TLC particles.*** (a) Steady-state heating; (b) transient heating.

2.3.5 Image Capturing

A digital still camera usually provides higher resolution than a digital camcorder. For steady-state measurement, the former is preferable. On the other hand, a camcorder that captures a series of still images with 25 frames per s in PAL or with 30 frames per s in NTSC is preferred since local temperatures on a target surface of interest change over in time if transient or unsteady flow behavior is considered. However, if the change in such a behavior is expected to be faster than either 20 ms, which is a typical response time of liquid crystals (Ireland et al., 1995), or the frame rate of a selected camcorder, the thermochromic liquid crystal technique is not suitable.

Example 2.2: Endwall Flow Pattern in a Sandwich Panel with a Truss Core

A. Background

A lattice-frame material (LFM) is one of the "periodic" cellular materials made possible by computer-based design and numerically controlled processing. The LFM consists of a three-dimensional (3D) network of cylindrical struts forming a tetrahedron as shown in Fig. 2.5. This tetrahedral unit cell is interconnected with other unit cells, which are periodically and spatially distributed. Recent developments in manufacturing techniques allow the strut diameter to vary from millimeters to tens of millimeters (Kim, 2004; Kim et al., 2004).

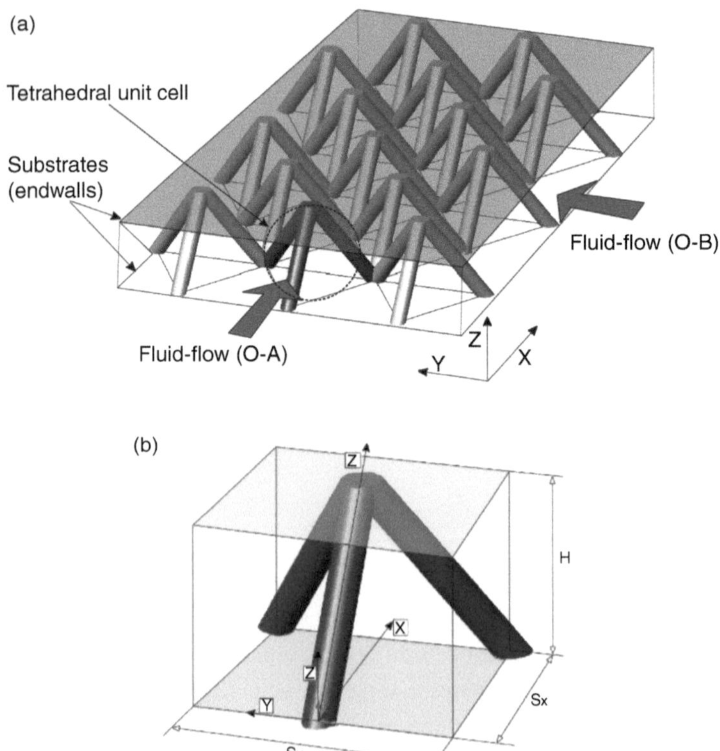

Figure 2.5 ***Three-dimensional (3D) topology of a lattice-frame material sandwiched between end-wall plates.*** (a) Sandwich panel; (b) unit cell.

B. Test Setup

B1. Test Rig

The experimental apparatus consists of four main sections: a contraction, a parallel flow channel, a test section, and an air supply system. Fig. 2.6a shows the test rig for the steady-state experiments, and Fig. 2.6b shows its modified version for the transient tests. Air at ambient conditions is drawn through the channel passage by a suction-type fan. The coolant enters through a 9:1 contraction followed by the parallel section before it reaches the test section. The parallel section length-to-channel height ratio is 20. A rectangular test section with dimensions of 0.127 m (width), 0.012 m (height), and 0.127 m (length) allows 86 tetrahedral unit cells to be accommodated in the test section.

The purpose of these experiments is to obtain detailed local variation of heat transfer on the inner surface of the sandwich cover plate. For transient heat transfer experiments with the TLC technique, the main sections of the test rig used in the steady-state experiments are modified to house

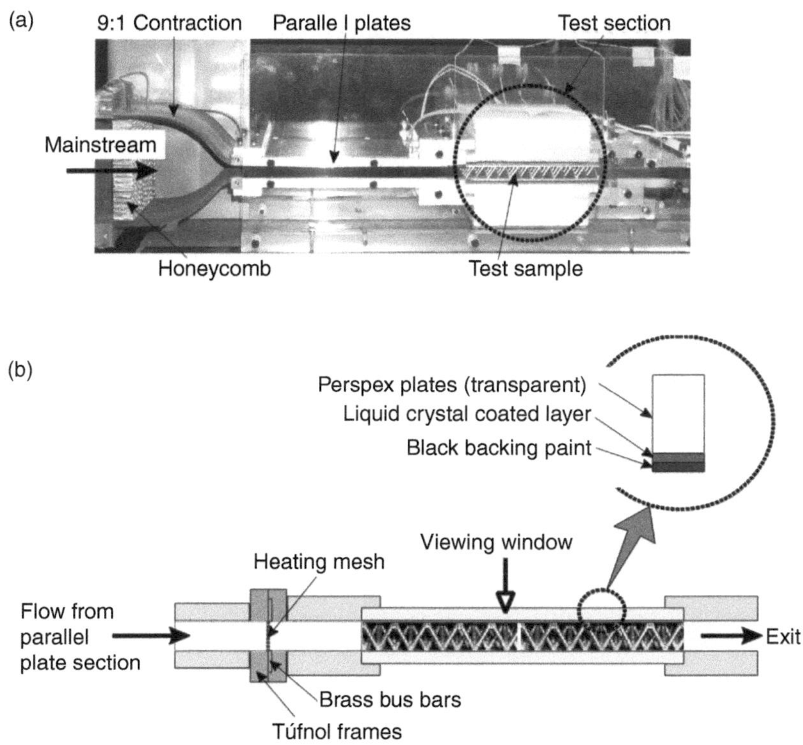

Figure 2.6 ***Photograph of a test rig for steady-state experiments and its modification for transient heating.*** (a) A steady-state test rig with 9:1 contraction and flow developing parallel channel before the test section; (b) a cartridge for transient measurement with a heating mesh and a TLC layer sprayed on an inner Perspex cover plate.

a cartridge-type transient test section as shown in Fig. 2.6b. The cartridge section contains a heating mesh, a parallel section, and a new test section. The heating mesh is used to provide a transient change of the fluid temperature. A parallel section of 87 mm length is placed between the heating mesh and the test section.

B2. Heating Mesh

During the data reduction process for the transient heat transfer experiments, 1D heat conduction across a semiinfinite solid (details discussed by Ireland et al., 1995) is assumed. In order to provide a temporal change of the fluid flow temperature, a heating mesh method is employed (as reported in Ireland et al., 1999; Wang et al., 1995, 1996, 1998).

The heating mesh is made of stainless steel wire of 36 μm diameter with 94 μm aperture (pitch of the wires), providing an open area ratio of 38%. It is sandwiched between two Túfnol frames after being soldered to brass bus bars. The bus bars are used to connect the heating mesh to a low-voltage but high-current transformer (used for arc welding). While air is passing through the heating mesh, the heat generated by the electrical current is transferred to the air. The steady-state increase of air temperature ΔT caused by the heating mesh can be estimated using:

$$\Delta T = \frac{I^2 R}{\dot{m} c_p} \tag{2.2}$$

where I, R, $\dot{m}$, and c_p are the imposed current, the electrical resistance of the heating mesh, the mass flow rate, and the thermal capacity of air, respectively. During the transient measurements, the air temperature rises exponentially due to the thermal capacity of the mesh. The fixed temperature rise ΔT can be obtained by adjusting either the coolant mass flow rate or the input current to the heating mesh. In the current example, the mass flow rate is fixed at a desired flow velocity, and then the current is adjusted to give the required ΔT.

B3. Test Model

The test sandwich model for the transient heat transfer experiments is constructed using a polycarbonate LFM core and two 6 mm thick Perspex cover plates. Before the Perspex plates were attached to the top of the core structure, TLC particles (BM/R27C6W/C17-10 from Hallcrest Inc.) were sprayed on the inner surface of one of the plates, which is used as a viewing window, followed by coating with a black backing paint layer (BB-G1 from Hallcrest Inc.). The test model is shown in Fig. 2.6b.

C. Endwall Surface Flow Patterns

C1. Orientation A

Fig. 2.7a and b display separately the endwall flow patterns (orientation A: Fig. 2.5a) on the bottom cover plate of the LFM sandwich obtained from oil flow visualization and TLC measurement techniques, both at

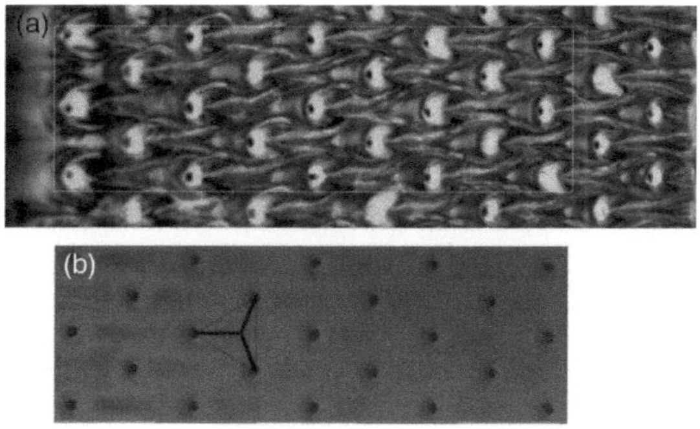

Figure 2.7 ***Comparison of endwall flow pattern of an LFM sandwich panel for Orientation A.*** (a) Oil-dye technique; (b) thermochromic liquid crystal (TLC).

$Re_{dp} = 8.0 \times 10^3$. The entry effect disappears after the second unit cell in the flow direction (along the x-axis). In the rest of the cells, similar endwall flow patterns over each cell are observed. The results also suggest that the flow is symmetric about a constant y-line passing through the center of each vertex of the LFM.

The endwall flow visualization images show very low momentum regions on the downstream side of the vertices on the endwalls. Higher flow interaction with the endwall is clearly observed in front of and behind the vertex. There is a saddle point, which is a singular point where zero shear stress exists, located in front of the stagnation point of the vertex. Note that the regions where a concentrated vorticity associated with arch-shaped vortices is situated correspond to the local minima of the endwall pressure distribution (Kim, 2004).

C2. Orientation B

Orientation B is rotated 90° from Orientation A (Fig. 2.5). Fig. 2.8 shows the endwall flow patterns obtained from both the oil flow visualization and the TLC measurements. In contrast to Orientation A, the entry effects persist up

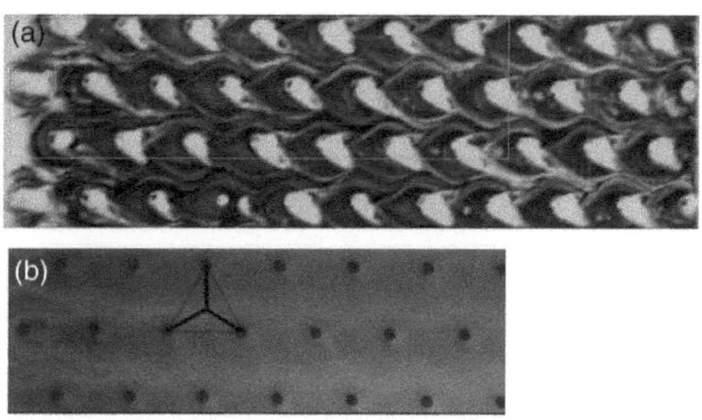

Figure 2.8 ***Comparison of endwall flow pattern of a lattice-frame material cored sandwich panel for Orientation B.*** (a) Oil-dye technique; (b) thermochromic liquid crystal.

to the fifth cell in Orientation B, highlighting a serpentine flow path located at 1/3 of the unit cell width. Furthermore, the axis of each horseshoe vortex is skewed with respect to the x-axis (in Fig. 2.5) at the center of each vertex due to the structural asymmetry of the LFM struts in the unit cell.

Comparing the endwall flow images of Orientation A with those of Orientation B, we can see similar features such as the formation of a horseshoe vortex at each vertex, the lower momentum regions at the downstream side of the vertices, and the relatively high flow interaction regions behind the vertices. However, the mainstream flow patterns in each unit cell and around each vertex are, in general, different for the two orientations.

2.4 SURFACE FLOW VISUALIZATION USING INFRARED THERMOGRAPHY

2.4.1 Background

Infrared (IR) light is a type of electromagnetic radiation that lies between the visible end of the light spectrum, and the microwave spectrum (Vollmer and Mollmann, 2010). Its wavelength ranges from about 800 nm to 1 mm and it is emitted particularly by heated objects. This radiant energy is often partially absorbed and reflected by bodies or particles, allowing it to be used by an optical tool such as an IR camera, which is called IR thermography. IR thermography is a technique in which a noncontact thermal device captures the infrared energy emitted by an object and converts it to temperature distribution images called thermograms. Thermograms represent different temperatures as various colors; for example, darker colors (deep blue, purple, and dark red) represent cooler temperatures, and lighter colors (bright red, orange, and yellow) correlate to warmer temperatures.

Although the fluid flow is at an average temperature, the fluid flow feature can be implicitly "seen" because of the shear stress that is present between a solid surface and the fluid flow immediately above it, which results in a difference in heat transfer (temperatures). However, this technique requires calibration to determine whether the readings from the IR camera are the same as those of the temperature transducer. Therefore, thermal sensors must be used in conjunction with IR thermography for the successful measurement of temperatures.

2.4.2 Surface Preparation

Temperature is calculated, by the IR camera, by measuring the IR radiation energy of an object and relating it to a blackbody that radiates an equivalent amount of energy according to Planck's Blackbody Law. However, there is an inherent problem in which the measured temperature, using the IR camera, may be different to the true temperature due to the object's different surface properties, which limit the radiation quantity. Emissivity and reflectivity are surface properties that must be considered for accurate temperature measurements.

Emissivity is a measure of efficiency of an object that radiates thermal energy. It is defined as the ratio of energy emitted from an object's surface to that radiated from a blackbody at the same temperature. A blackbody is a theoretical object or a system in which all electromagnetic radiation is absorbed and is a perfect emitter of thermal energy with an emissivity value of 1. The emissivity of an object, therefore affects how much energy an object emits and this information must be known for accurate temperature measurement, for example, if an object has an emissivity value of 0.77 (anodized aluminum) and its constant true temperature is 50°C, then the apparent temperature measured by the IR camera will be 38.5°C. Therefore, the emissivity values of the object being measured must be defined in the settings of the IR camera when taking measurements as the camera cannot automatically detect the emissivity of objects. This will ensure the temperature measurement is as close to the true temperature measurement as possible.

Reflectivity is the percentage at which electromagnetic radiation is reflected at a given wavelength by an object due to its physical properties. This mostly occurs from materials that are highly reflective such as polished aluminum, which reflects a large amount of radiation coming from the ambient. This is a problem as this will not provide an accurate temperature measurement for the IR camera, but it can be solved by adding a thin layer of thermally black paint to an object or a surface that has the property of high reflectivity. It must be noted that colors of objects or surfaces have no major effects on the IR temperature measurements and can be ignored unless the source of the color is a radically different substance to the main body of material because the emissivity and reflectivity will change, for example, metallic paints that contain aluminum (Merchant, 2012).

2.4.3 Calibration

Like a digital camera, an IR camera is susceptible to perspective distortions from different viewing angles and distances. This results in an incorrect temperature measurement that adds errors to an experiment, therefore it is desirable to calibrate the IR camera in place (in situ). Hence, the viewing angle and distance must be taken into consideration for accurate temperature measurements (Usamentiaga et al., 2014).

2.4.3.1 Viewing Angle

It is desirable to position the IR camera perpendicular to an object to ensure an accurate temperature measurement. However, if due to the nature of the experiment taking a perpendicular image is difficult or impossible, a thermocouple can be used to calibrate the temperature measurement of the IR camera. A thermocouple can be placed on a reference point on an object or surface and matched to the IR camera temperature at the same location by adjusting the emissivity setting on the IR camera.

2.4.3.2 Viewing Distance

The distance from the IR camera to the object which is being photographed affects the temperature measurement of the experiment. This is because the resolution of the image changes depending on the distance at which the object or surface is photographed. The further away the object is, the less area there is available on the IR camera's sensor to resolve the object's temperature measurement. If the object is too close, there is a risk that the vital temperature measurement of the object is left out. Therefore, there needs to be a balance between the two factors for optimum temperature measurements.

The distance parameter also needs to be defined in the settings of the camera to account for the change in the object's emissivity and reflectivity due to the change in distance.

Amongst other parameters, the object emissivity, reflectivity, ambient temperature, and relative humidity are required to be set on the IR camera for accurate temperature measurements.

Example 2.3: Round Jet Impingement on a Flat Plate

A. Background

Impinging jet cooling (or heating) has long been studied due to its superior heat transfer capability to other convective heat transfer schemes. A circular impinging jet on a flat plate serves as a fundamental configuration amongst others. Consequently, a multitude of studies have been devoted to this configuration (Gardon and Akfirat, 1965; Gauntner et al., 1970; Martin, 1977; Jambunathan et al., 1992; Reungoat et al., 2007).

Approaching the flat plate, the round jet is decelerated and it eventually stagnates on the flat plate, followed by radial deflection as shown in Fig. 2.9a. It has been argued (Yang et al., 1999; Gauntner et al., 1970;

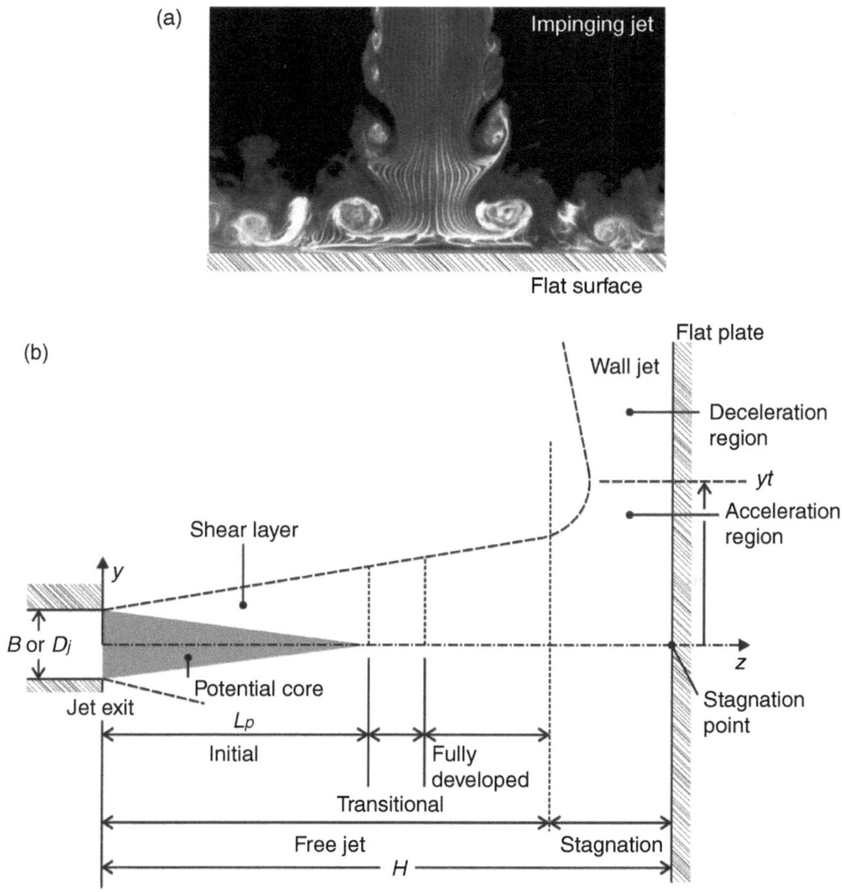

Figure 2.9 ***Flow pattern and description of a jet impinging on a flat plate.*** (a) Instantaneous flow pattern of a round jet impinging on a flat plate; (b) schematic of the time-averaged flow patterns. *Source: (a) (www.tvt.kit.edu).*

Hofmann et al., 2007) that free jet structures affected by the target flat plate are limited within approximately $2.0D_j$ upstream, forming the so-called "deflection zone," and that other flow regions (such as initial, transitional, and fully developed regions) remain unaffected, as illustrated in Fig. 2.9b. Inside the deflection zone, a wall jet region is formed. After stagnation, the jet is radially (or laterally) deflected and undergoes contraction and acceleration, followed by a deceleration.

A laminar boundary layer develops from the stagnation point and its thickness grows. It should be noted that regardless of the jet conditions (ie, either turbulent or laminar jet) approaching the flat plate, a laminar boundary layer always starts to develop. The wall parallel velocity component initially increases, reaching a maximum value at a certain distance y_t

(Fig. 2.9b) from the stagnation point. The stabilizing effect of acceleration keeps the boundary layer laminar in the stagnation zone, while transition to turbulence generally will occur immediately after $y = y_t$ in the deceleration region (Martin, 1977).

This example shows the local axis-symmetrical distribution of flow features on a flat plate posed by impingement of a round turbulent jet using IR thermography.

B. Experimental Details

Fig. 2.10 shows a schematic of the present test rig consisting of a round jet impinging on a flat plate (0.5 m × 0.5 m). The flat plate was fabricated from a 10 mm thick acrylic material ($k \sim 0.2$ W/mK). For heat transfer measurement, a circular film heating element ($D = 0.3$ m with thickness of 0.27 mm) was attached to the target surface, emitting a constant heat flux. Two foil and one bead-type precalibrated T-type thermocouples were used to measure the temperature at the stagnation point on the flat plate, on the back of the target surface (to estimate heat loss), and at the round jet exit, respectively.

Air at ambient conditions drawn by an axial fan controlled via an inverter was discharged from the round jet nozzle with a fixed diameter of $D_j = 0.026$ m. The round jet nozzle was mounted on a linear traverse system so that the jet exit to target surface distance, H, along the jet axis could be varied systematically.

The velocity of the round jet along the centerline axis (ie, the z-axis) was measured to quantify the length of potential core with a Pitot tube in conjunction with a digital pocket micro-manometer (DPM ST650™). The

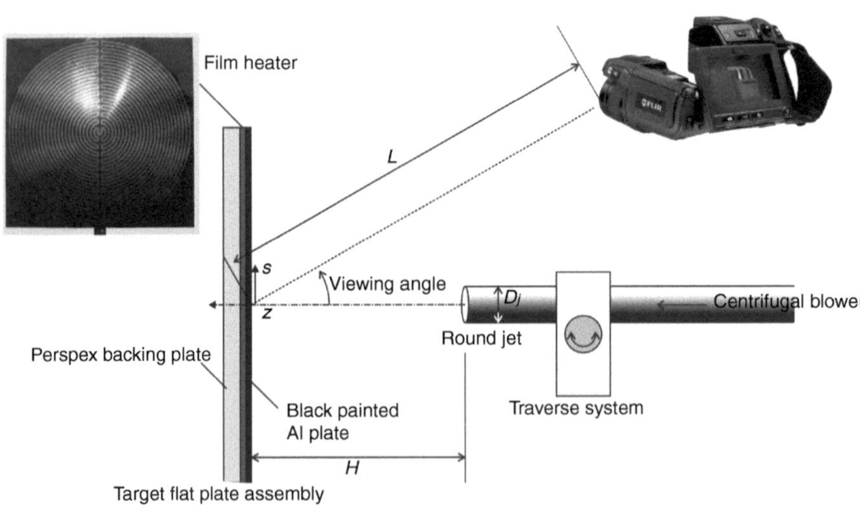

Figure 2.10 ***Schematic of the test setup for round jet impingement cooling on a uniformly heated flat plate.***

Pitot tube was linearly traversed along the centerline axis of the jet in free exit and in impingement on the flat plate.

The jet Reynolds number based on the round jet diameter (D_j) and jet centerline exit velocity (w_e) is defined as:

$$\mathrm{Re}_{Dj} = \frac{\rho w_e D_j}{\mu} \tag{2.3}$$

During the entire experiment, Re_{Dj} was fixed at 20,000. The heat transfer characteristics were evaluated using the Nusselt number, defined based on the round jet diameter D_j as:

$$\mathrm{Nu}_{Dj} = \frac{h_s}{k_f / D_j} \tag{2.4}$$

where

$$h_s = \frac{q''_{net}}{T_s - T_j} \tag{2.5}$$

Here, q''_{net} is the net heat flux, k_f is the thermal conductivity of air, whilst T_s and T_j are the local surface temperature (or T_{IR}) and jet temperature at the jet exit, $z = 0$, respectively. The net heat flux, q''_{net}, is obtained through energy balance:

$$q''_{net} = q''_t - q''_{cond} - q''_{rad} \tag{2.6}$$

where q''_t is the total heat flux emitted from the film heating element, and q''_{cond} and q''_{rad} are the conduction and the radiation heat loss, respectively. The conduction heat loss through the backside of the target surface was estimated using Fourier's law and found to be 3.5%. The radiation loss is calculated using the Stefan–Boltzmann law:

$$q''_{rad} = \varepsilon\sigma\left(T_s^4 - T_a^4\right) \tag{2.7}$$

Here, ε is the emissivity, σ is the Stefan–Boltzmann constant, and T_a is the ambient temperature. The emissivity was taken to be 0.94 as estimated by an infrared camera (FLIR T-Series).

The resolution of the temperature readings from the temperature scanner was ±0.1 K for each thermocouple. The errors associated with the multimeter readings for the input voltage to, and the electrical resistance of, the film heater were estimated to be ±0.09 V and ±0.27 Ω, respectively, resulting in a total heat flux error of ±11.1 W/m^2. The uncertainty of the net heat flux was estimated to be within 1.1%. Therefore, the uncertainty of the Nusselt number was calculated to be within 3.3%.

Despite the benefits of IR thermography, it also has limitations in the form of many parameters which can cause measurement error; this includes parameters to consider during calibration (object emissivity, reflective temperature, ambient temperature, relative humidity, distance). For the current study, an IR camera (FLIR T-Series) was used to measure the local heat transfer distribution on a target surface, subjected to an impinging round jet. The emissivity was calibrated using the contact method described in Usamentiaga et al. (2014) as follows:

- The target surface was heated without the jet, until it reached temperatures similar to that during jet impingement.
- A film thermocouple flush mounted on the target surface [at the stagnation point (Fig. 2.10)] was used to obtain a reference temperature.
- Once the target surface had reached temperatures similar to that during jet impingement, a reference temperature was measured with the thermocouple, then the IR camera emissivity was changed until it detected a temperature matching with the reference temperature at the same location.

A second IR camera emissivity calibration method called the reference emissivity material method is described in Usamentiaga et al. (2014) and can also be used.

The reflected temperature was estimated using a method also described in Usamentiaga et al. (2014) and ASTM E1862-97(2002)e1 (1997) as follows:

- A crumpled and re-flattened piece of aluminum foil was used as a reflector.
- The re-flattened aluminum foil was positioned in the field of view of the IR camera.
- The temperature of the foil was measured with the emissivity and distance set at 1 and 0 on the camera.
- The reflected temperature was measured by applying the temperature of the foil to the surface; the resulting temperature value is the reflected temperature.

Amongst other parameters causing a measurement error while using an IR camera, the view angle is commonly encountered due to the fact that it is not always possible to align the camera with the surface to be measured. To account for the effect of the view angle, the temperature distribution on the target surface was captured with the IR camera for the isothermal case (without an impinging jet). IR images were captured while constant uniform heat flux (10, 25, and 30 W/m^2) was maintained on the target surface. The IR images were processed using FLIR Tools+ software in order to extract the temperature distribution along the *s*-axis, and compared to the thermocouple [at the stagnation point (Fig. 2.10)] measured temperature. The temperature of the thermocouple at 0° relative viewing angle was matched to the thermocouple temperature at the same location by adjusting the emissivity setting on the IR camera.

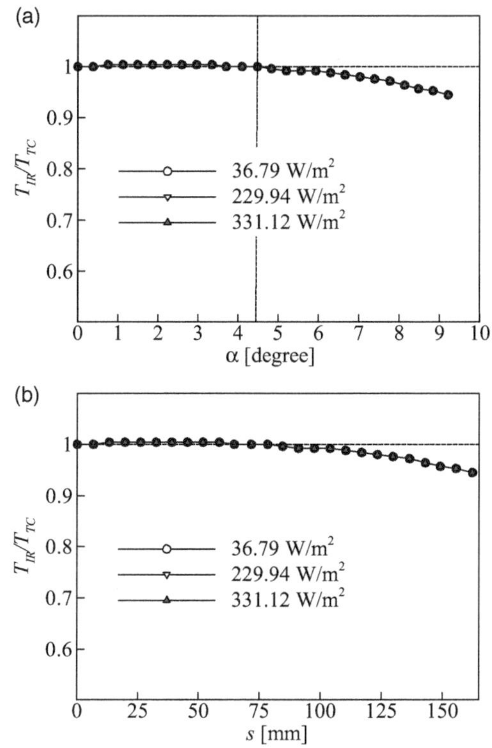

Figure 2.11 ***Effect of viewing angle on IR temperature normalized by thermocouple temperature (T_{IR}/T_{TC}) on a flat plate versus: (a) the IR camera relative view angle (α); (b) the radial location (s).***

Fig. 2.11a and b show that when the camera (lens) is positioned 1.0 m away from the flat plate (which is the actual distance set up), the deviation of the temperature occurs at a relative viewing angle of approximately 4.45° for the flat plate, regardless of the applied heat flux values. To correct errors associated with the view angle, the curves in Fig. 2.11 were used to derive a fourth-order polynomial expression of the curve trend, which was used to correct errors of IR temperature associated with the viewing angle.

C. Discussion of Results

C1. Length of Potential Core for the Free Round Jet

This study considers a round unconfined jet discharged to a surrounding fluid (of the same fluid properties), that is, a submerged jet, removing heat from a flat plate which emits uniform heat flux. Therefore, it is instructive to establish the free round jet characteristics.

According to the classical description of free jet structures (Abramovich, 1963; Gauntner et al., 1970; Viskanta, 1993), three distinctive flow regions exist: (1) the initial region, (2) the transitional region, and (3) the fully

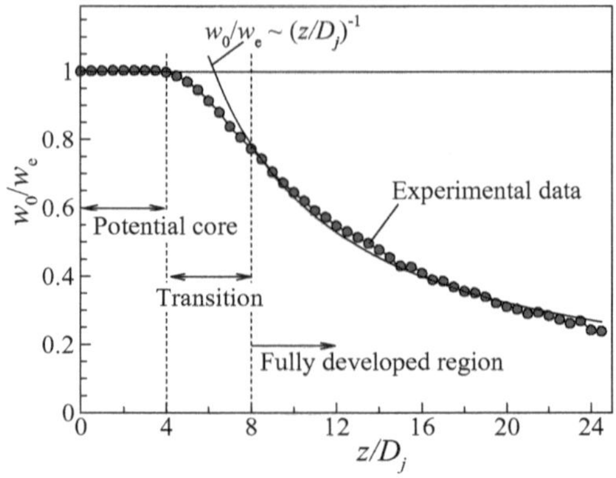

Figure 2.12 ***Free jet characteristics at Re_j = 20,000 showing axial variation of normalized centerline velocity, w_0/w_e, indicating a potential core persisting up to z/D_j = 4.0.***

developed region. In the initial region, the jet velocity along the jet centerline axis maintains its magnitude as high as that at the jet exit. In the transitional region the jet centerline velocity begins to decrease substantially, transitioning to the fully developed region where it decays. Furthermore, a region where the jet is undisturbed by the entrainment is present, the so-called "potential core." After the jet is fully developed, the axial velocity profiles become self-similar, collapsing onto a single curve that could be fitted using Gaussian distribution (Gauntner et al., 1970; Viskanta, 1993).

Following the definition of Giralt et al. (1977), the potential core length is the distance from the jet exit to a downstream location where the centerline velocity w_0 reaches 98% of the jet exit velocity w_e. It is derived from Fig. 2.12 that the present free slot jet has the length of potential core, z/D_j = 4.0.

C2. Temperature Map and Heat Transfer Distribution

Heat transfer distribution on a flat plate that is cooled by a round jet is axis-symmetric with respect to the jet axis (Fig. 2.13). The jet is spread radially for the round jet after impingement. Therefore, the momentum is reduced while convecting away from a stagnation point on a flat plate coinciding with the round jet axis. The jet then experiences an increase in flow area, which leads to a decrease in the local Reynolds number, resulting in the radial decrease in the local heat transfer.

The highest heat transfer on a flat plate typically occurs at the stagnation point, whose magnitude varies with jet exit to flat plate distance (H) at low turbulence levels (Gardon and Akfirat, 1965) and low Reynolds numbers (Hofmann et al., 2007). In other words, for a round jet removing heat from the flat plate, the lowest temperature exists at the stagnation point.

Depending on the axial position relative to the free jet's length of inviscid core, the so-called "potential core," which has been characterized for

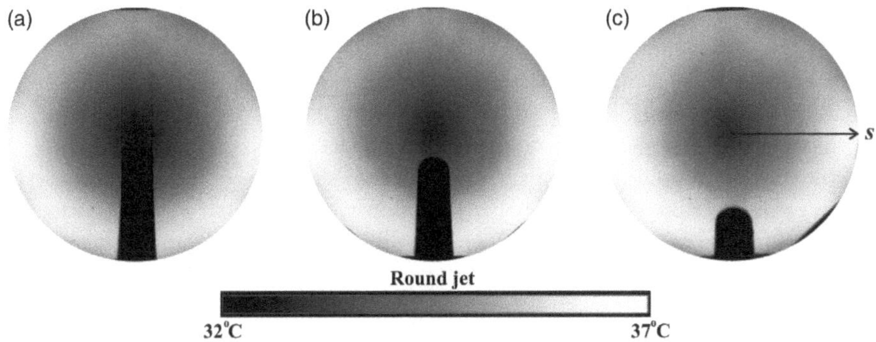

Figure 2.13 ***Temperature (T_{IR}) maps on a flat plate cooled by a round impinging jet at Re_{Dj} = 20,000.*** (a) $H/D_j = 2.0$; (b) $H/D_j = 4.0$; (c) $H/D_j = 8.0$.

the present round jet in Fig. 2.12, either two peaks of local heat transfer (a primary peak at the stagnation point and a second peak away from the stagnation point) or a single peak (only at the stagnation point) exist on a flat plate. The appearance and disappearance of the second peak are associated with the laminar to turbulent transition of the wall jet formed on the flat plate as a result of the lateral deflection of the jet.

The radial distribution of the calculated convective heat transfer based on the extracted temperature, T_{IR} (in Fig. 2.13), on the flat plate positioned inside the potential core ($H/Dj = 2.0$) is plotted in Fig. 2.14. Local heat transfer distribution as a Nusselt number is defined in Eq. (2.4) as: $\mathrm{Nu}_D = q''_{net}/(T_{IR} - T_j)$ where q''_{net} is the net heat flux removed by the round jet and T_j is the jet temperature measured by a bead-type thermocouple and shows its decreasing from the stagnation point, reaching a local minimum at $s/D_j = 1.5$. Afterwards, it increases, peaking at $H/D_j = 2.0$, followed by a monotonic decrease. These peaks are attributable to the transition from

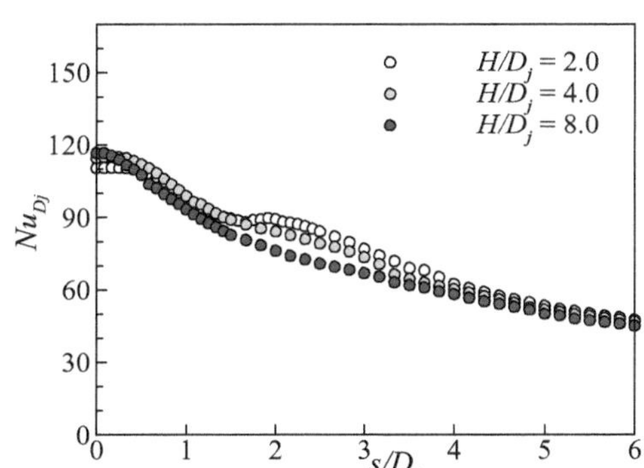

Figure 2.14 ***Distribution of heat transfer on a flat plate for selected jet exit to flat plate distances (H/D_j) at Re_{Dj} = 20,000.***

laminar to turbulent flow (Jambunathan et al., 1992; Gauntner et al., 1970; Gardon and Akfirat, 1965).

When the flat plate is positioned near the tip of the potential core ($H/D_j = 4.0$) and outside the potential core ($H/D_j = 8.0$), the second peak in heat transfer disappears, showing a monotonic decrease in heat transfer from the stagnation point, which may indicate either that transition from laminar to turbulent flow is absent or that transition takes place but too early and near the stagnation point, therefore it could not be distinguished.

2.5 IN-FLOW VISUALIZATION USING NEUTRALLY BUOYANT HELIUM BUBBLES

2.5.1 Background

To visualize flow pattern, bubbles produced in a helium bubble generator can be used. The neutrally buoyant bubbles produce clearer streamlines by being more distinct than smoke or solid particulates. Bubbles have very low inertial mass, thus allowing the bubbles to indicate flow features accurately. The advantage of using bubbles is that a gas containing bubbles may be filmed passing through fans, straightening grids, and a honeycomb with minimal bubble loss. Individual bubbles can be tracked to indicate streamlines.

The low density and inert properties of helium gas make it possible to create approximately neutrally buoyant bubbles for flow visualization. The helium bubbles consist of gaseous helium enclosed in a dense liquid film. The generation of the bubbles is relatively simple and is achieved in a mixing nozzle. The nozzle consists of three annular tubes. Helium gas flows through the small central tube. Bubble film solution (BFS) is sucked by the helium into a larger tube that surrounds the tube with helium at a specified flow rate. The helium gas in the center tube blows through the bubble film solution to create a stream of resilient bubbles at the end of the film solution tube. Low-pressure dry compressed air is blown through the outer tube to transport the bubbles.

The generators separate bubbles that are too large or too small to be neutrally buoyant in a cyclone filter. The neutrally buoyant bubbles may be transported in a flexible tube to where they are required for flow visualization. Approximately 300–400 bubbles per second can be generated (SAI, 2014). The bubble diameter ranges between 0.64 mm (0.05″) and 3.8 mm (0.15″) using helium at 1.4 bar with a flow rate of approximately 0.2 L/min. For most applications, the regulated pressure of the dry compressed air may be 2–4 bar with a flow rate of 34 L/min. The bubble film solution is a proprietary composition, and consumption is typically 50 cm^3/h (SAI, 2014).

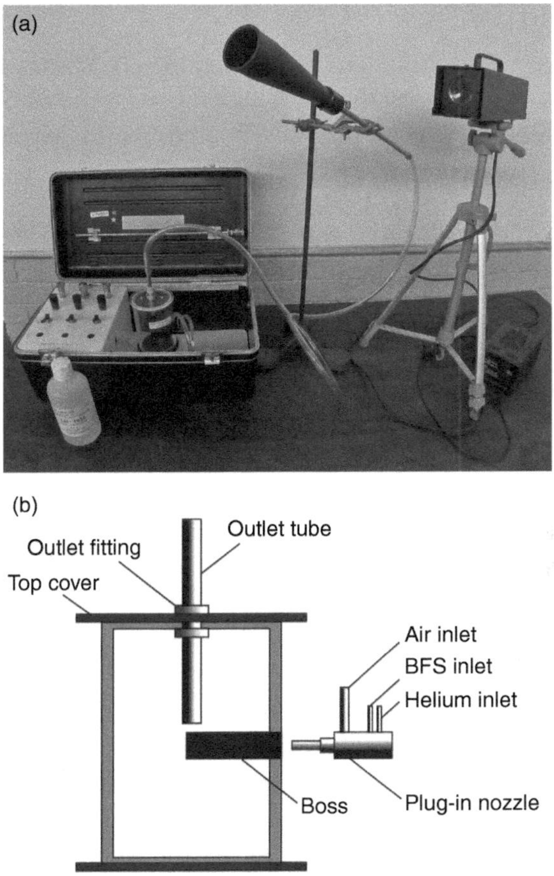

Figure 2.15 ***A helium bubble generator with light source.*** (a) Photograph of a complete unit; (b) schematic of the cyclone filter.

The assembled helium bubble generator is shown in Fig. 2.15a. A schematic of the cyclone filter is given in Fig. 2.15b. The nozzle head plugs into a boss on the outer surface of the vortex filter. The bubble-laden air stream is introduced into the filter tangentially. The heavier bubbles enter the vortex within the filter and sink, creating a small amount of film solution on the base of the filter. The lighter bubbles enter the filter and rise towards the top cover. They pop on contact and a small amount of the film solution accumulates on the cover. The number of bubbles popping against the top cover is a good indication of the mixture between BFS and helium when adjusting bubble quantity with the metering valves. Neutrally buoyant bubbles enter the vortex within the filter and move towards the center of the filter where they are blown out of the outlet tube.

2.5.2 Seeding the Flow

The bubble-laden air mixture can be injected into an airstream with velocities of less than 60 m/s using the apparatus shown in Fig. 2.15. The velocity of the bubble mixture may be throttled by adjusting the air metering valve. The application would generally determine any limitations to the introduction of bubbles. As an example, extensive visualization of low-velocity air jets has been achieved using the helium bubble technique. The air jets were produced using a blower, ducting, and an air-straightening exit nozzle. Two examples of flow seeding are given as follows.

For the first example, the interaction of an air jet and a surface needed to be visualized. Both flat and concave surfaces were investigated. This could be defined as macro flow visualization. A top view of the experiment is shown in Fig. 2.16. The bubbles were introduced into the duct through the wand upstream of the slot jet. The bubbles were introduced perpendicular to the air flow. This seems counterintuitive, but the bubble distribution was better when compared with parallel introduction of bubbles. The bubbles need to have sufficient time to accelerate to the flow velocity before entering the nozzle. This was done because the entire flow jet was being visualized. The flow characteristics could then be visualized effectively.

Fig. 2.17 illustrates a typical raw photograph of an experiment. The bubble paths are visible at a series of locally parallel lines due to the bubbles reflecting light at their tangents in the direction of the camera. Glare is frequently a problem. Even with a matte black surface, the powerful light source produces a significant amount of reflected light. The light can also be seen reflecting off the slot jet and concave surface. With careful editing, glare can be removed. The photograph properties with a Nikon D5100 were *f*

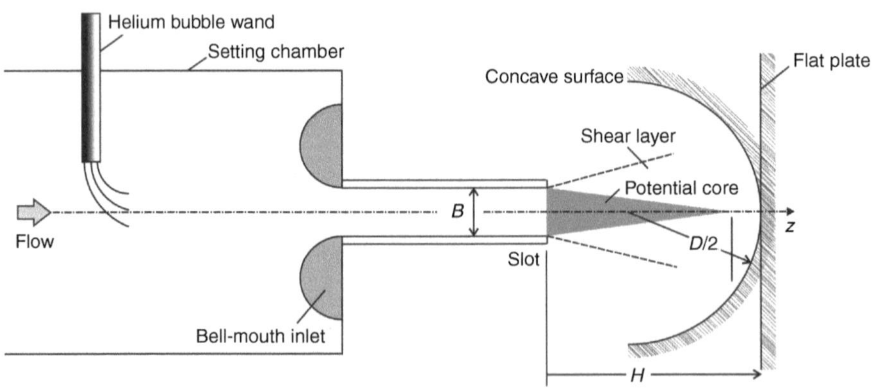

Figure 2.16 ***Schematic of the arrangement for perpendicular bubble introduction for macro flow visualization.***

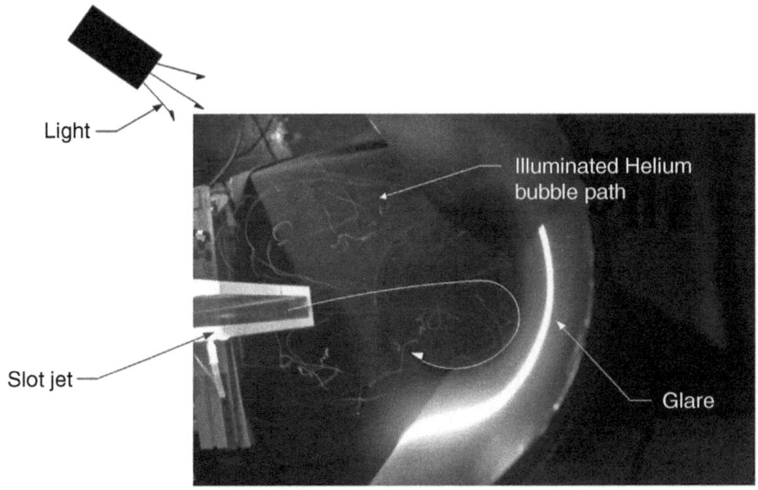

Figure 2.17 ***Unprocessed image of macro 2D flow (slot jet).***

3.5, 6 s exposure time, focal length 18 mm, and ISO 100 (see section 2.5.4 for explanations of these terms).

Parallel introduction of bubbles is best suited to situations where visualization of a micro flow feature within the air stream is required. This will result in a narrow stream of bubbles aligned with the flow with denser bubble distribution for better photographic results. A typical example of the arrangement is shown in Fig. 2.18.

2.5.3 Illumination

For flow visualization, the bubbles must be injected into the flow upstream of the flow feature being investigated. This makes the visualization of 3D

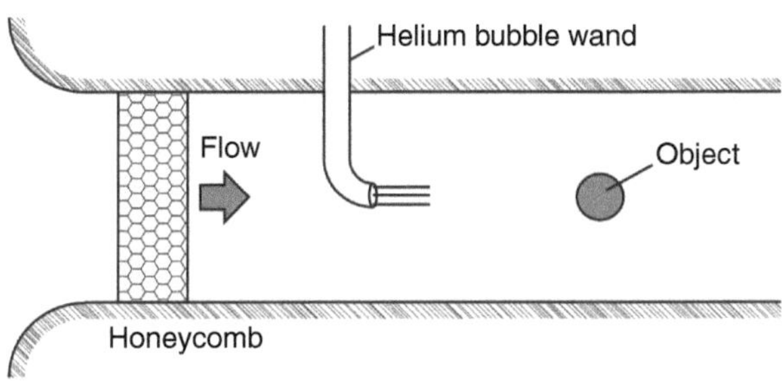

Figure 2.18 ***Schematic of the arrangement for parallel bubble introduction for micro flow visualization.***

flow features fairly simple. The bubbles trace streamlines which, depending on the turbulence intensity, can make illumination of the bubble streamlines challenging. The 3D case will be dealt with first. Any powerful light source may be used. The light source needs to be capable of conical beam illumination and light-slit illumination. Most light sources may be modified using lenses and knife-edges to vary the aperture of the slit. The light may be beamed with the flow direction, against the flow direction, and perpendicular to the flow direction.

To illuminate a 3D flow feature, the direction of the streamlines will define the lighting position. An example is given in Fig. 2.19a and b where wingtip vortices from a NACA0012 airfoil were investigated. The airfoil had a chord of 300 mm and was filmed in a draw-down wind tunnel with air at a velocity of 10 m/s. The light source was placed aft of the wingtip. The aluminum wing was darkened in the editing of the photograph. The photograph properties with a Canon PowerShot G9 were *f* 4.5, 1 s exposure time, focal length 29 mm, and ISO 80.

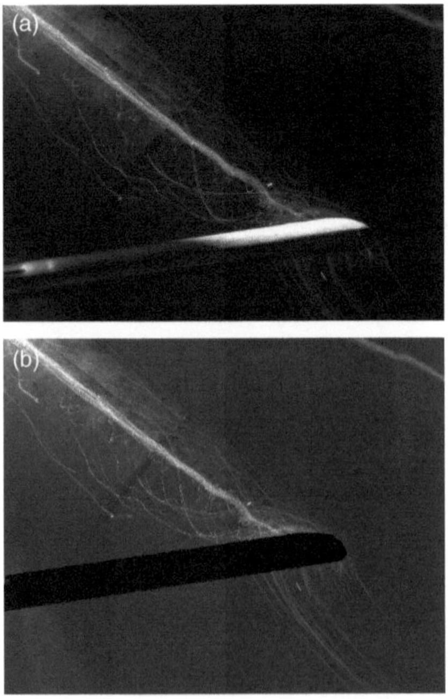

Figure 2.19 ***Wingtip vortex from a NACA0012 airfoil at 10 m/s.*** (a) Correct exposure of bubbles; (b) same image with a wing removed and background contrast manipulated to highlight vortex.

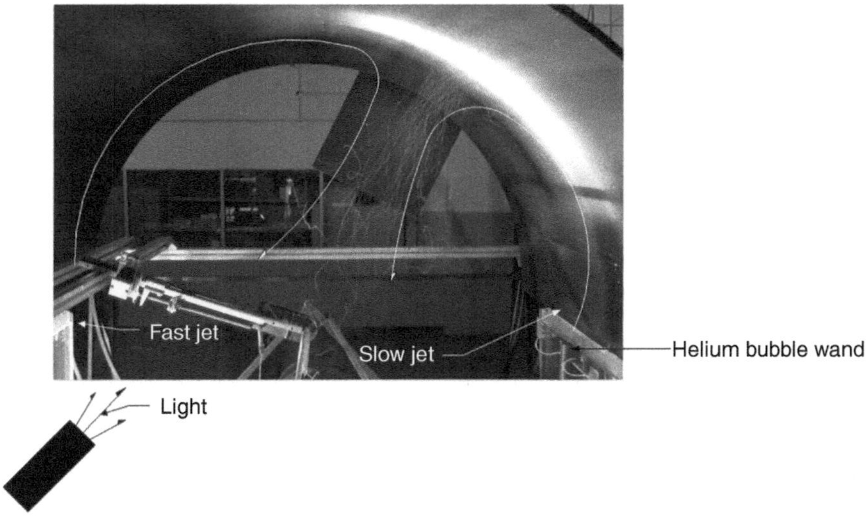

Figure 2.20 ***Unprocessed image of macro 2D flow.***

Illumination of the bubbles within the flow is fairly difficult for 2D flow features. The example in Fig. 2.20 is a large-scale concave surface with a tangential slot air jet on opposite sides of the concave surface. The angle of the converging slot jet relative to the local tangent needed to be determined. The two slot jets had differential air velocities for the test case shown. The raw photograph is used to illustrate the difficulty in eliminating glare. A light-slit technique was used to capture a 2D slice of the flow in the center of the concave surface. The helium bubbles were introduced using the parallel introduction technique next to the slower slot jet to increase bubble density in the photograph. The photograph properties with a Nikon D5100 were *f* 4.8, 3 s exposure time, focal length 20 mm, and ISO 800.

2.5.4 Photographing

Digital cameras have made the photography of flow fields fairly easy. The lighting, camera positioning, and editing are most important to capturing good-quality images of flow features. The basic camera equipment required is a tripod and digital single lens reflex (DSLR) camera. The majority of flow visualization is carried out in low-light high-contrast environments. Therefore, the best results are obtained using manual photographic mode on the camera. Manual mode allows for light sensitivity, white balance, aperture, and shutter speed to be controlled by the user.

The light sensitivity of the sensor, referred to as ISO, may be manually changed to change the exposure time. Camera automatic mode generally uses high ISO, which can make the photograph grainy. White balance allows for correct color in photographs, which is not important for most flow visualization applications and can be corrected in the editing process if necessary. The camera lens aperture may be adjusted in manual mode on most cameras. The aperture scale, referred to as *f*, restricts the amount of light passing through the lens. For low-light conditions, automatic camera mode uses a low *f* scale (wide aperture) to allow in maximum light. This has the negative effect of reducing the depth of field. In manual mode, a higher *f* scale (small aperture) allows in less light, giving the photograph greater depth of field. Greater depth of field improves the clarity of the bubble streamlines.

Shutter speed is set by the user in manual mode and is shown in seconds for low-light conditions and fractions of a second for normal photography. The use of a camera in manual mode means the user needs to understand the aperture, ISO, and shutter speed and the impact each has on the exposure. The shutter speed is most important when capturing bubble streamlines. The shutter speed defines the streamline length that a bubble will make on a photograph. As an example, for air moving at 10 m/s, a shutter speed of 1/250 s will result in a streamline length of 40 mm. This may be too short to indicate a flow feature. The exposure of a photograph may be determined before the photograph is taken since all cameras have an exposure indicator. Underexposure indicates that the photograph will be underexposed with very dark features. Overexposure indicates that the photograph will have very light features. Two exposure cases are shown in Fig. 2.21a and b. The aperture setting makes a large difference to the background light in the photographs.

The photographs in Fig. 2.22a and b were taken with the same camera settings and indicate the effect that camera position has on an image. The left image shows the flow field with more detail and less glare than does the right image. The photograph properties with a Nikon D5100 were *f* 4.8, 3 s exposure, focal length 20 mm, and ISO 800.

It is recommended that the camera be operated in automatic focus mode while the photographing is set up. The camera may be focused on a focus target placed in the center of the illuminated flow field. The camera should then be switched to manual focus to prevent the focus from changing during photographing. The automatic focus function will keep changing focus as the bubbles move. Fig. 2.23 indicates a vortex filmed in perfect conditions. The vortex was illuminated behind the camera position. The photograph

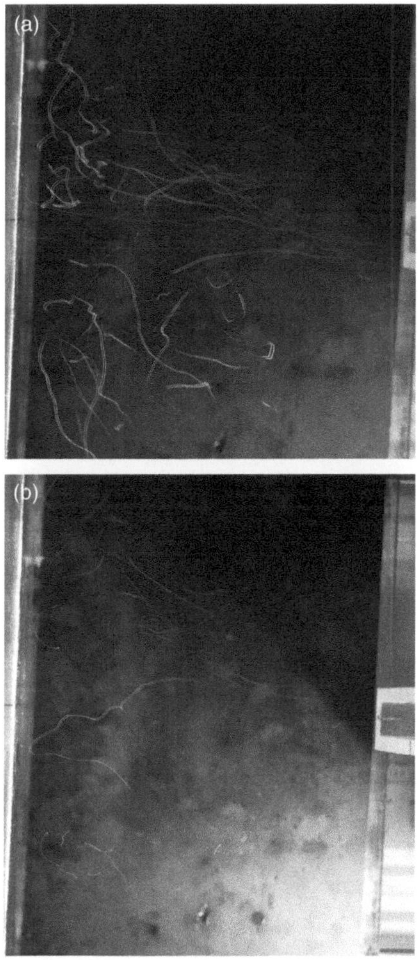

Figure 2.21 ***Indication of the effect of aperture on exposure.*** (a) Correct exposure with Nikon D5100, focal length = 34 mm, *f* 5, exposure time = 1 s, ISO 2000; (b) overexposure with Nikon D5100, focal length = 30 mm, *f* 4.5, exposure time = 1 s, ISO 2000.

captured an oblique view of the vortex to eliminate glare and reflection, and the photograph is unedited. The photograph properties with a Canon EOS 400D were *f* 5.6, 1/5 s exposure time, focal length 53 mm, and ISO 1600.

Depending on the purpose of the flow visualization exercise, several editing options exist to achieve the required results. Many software packages are available to edit photographs. An example will be given using the GNU image manipulation program (GIMP). The threshold function was found to be a very powerful tool to eliminate glare, background, and reflection. The threshold function converts the image into black and white pixels. The

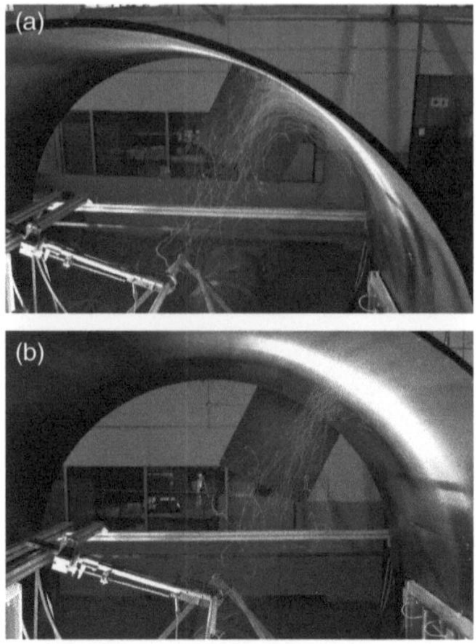

Figure 2.22 ***Indication of the effect of camera position on image quality.*** (a) Unprocessed photograph with reduced glare; (b) unprocessed photograph with glare.

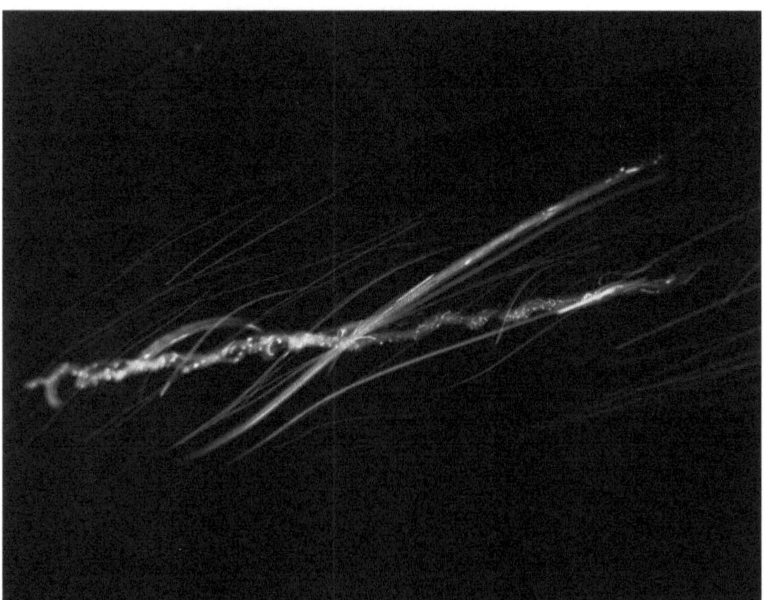

Figure 2.23 ***Illuminated 3D vortex.***

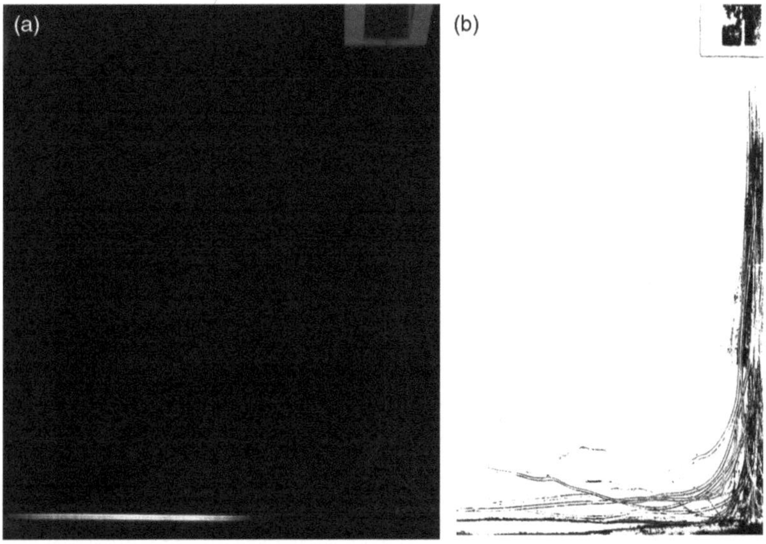

Figure 2.24 ***Example of poor-quality photograph editing.*** (a) Original image; (b) edited image.

threshold function then allows the user to eliminate pixels that do not fall within the threshold range. Due to the contrast of the white bubbles against a grey to black background, the function allowed for good images to be achieved with little effort. Fig. 2.24 shows a raw photograph (Fig. 2.24a) and the same photograph after the threshold levels were adjusted to eliminate everything except the streamlines (Fig. 2.24b). The photograph was then inverted to turn it into a negative. This is an example of a poor photograph edited to indicate flow features.

2.6 IN-FLOW VISUALIZATION USING PLANAR LASER IMAGING

2.6.1 Background

Planar laser imaging is another method that may be employed to reveal a particular flow structure that is of interest to the investigator. It involves illuminating a thin "slice" of the flow field using a planar laser light sheet that has been created from special optics. The Particle Image Velocimetry (PIV) system may be used for this purpose. A typical PIV system consists of five key components:

1. Flow test rig (such as a wind tunnel),
2. Seeding generator (used typically for seeding gaseous flows such as air),

3. Pulsed laser system with light sheet optics, which creates a planar light sheet,
4. Charge Coupled Device (CCD) digital camera system,
5. Data acquisition and processing software.

Further discussion on how to set up and implement the PIV system, to measure a particular velocity field, is provided in Chapter 5. For the present, we consider the PIV system as an integrated system, where a high-energy pulsed laser light source is synchronized with the camera exposure to capture instantaneous flow field images, that is, *flow visualization*. To visualize the flow field, seeding particles are introduced and carried by the flow. An underlying assumption of the PIV technique is based on the ability of the particles to closely follow the flow field, so that the motion of the particles is a close representation of the actual structure of the flow. Hence, the size and material properties (eg, density) of the seeder particles in relation to the working fluid of the experiment are an important consideration. In addition, the PIV data acquisition and processing software is a convenient way to store an ensemble of raw flow images, which can be processed further.

The general layout of the PIV system is shown in Fig. 2.25. A key feature of this method is the orthogonal positioning of the camera in relation to the pulsed laser light sheet, so that the optical axis of the camera is perpendicular to the plane of the laser-thin laser light sheet. The camera then captures instantaneous images of the particles that are illuminated by the pulsed laser. The pulse duration is very short (ie, ≈ 10 ns), which causes a stroboscopic

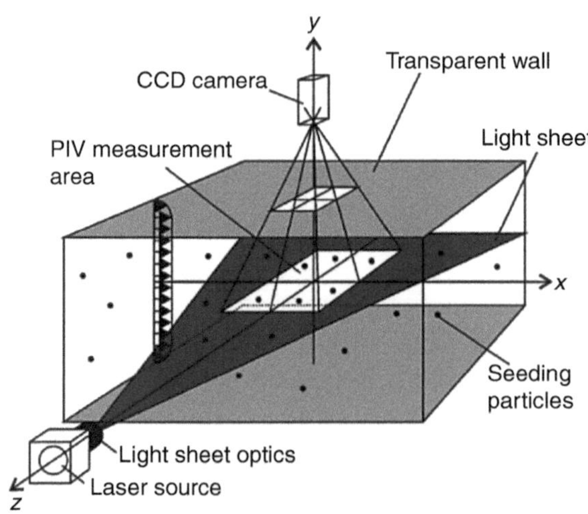

Figure 2.25 ***Typical 2D PIV setup, with an external planar light sheet, that illuminates the seeding particles that are in the plane of the light sheet.***

effect where moving particles appear stationary during illumination (Raffel et al., 1998). This means the images of the particles that are carried by the flow field appear clearly defined which allows certain flow structures such as eddies to be made visible.

A difference between implementing the PIV system for velocity field measurements versus flow visualization is related to how the seeding particles are dispersed into the flow field that is being investigated. For PIV measurements it is desirable to have a high concentration of seeding particles that have a *homogeneous* distribution in the flow field; the effect is that flow structures are *not* directly visible from the raw images. On the other hand, flow visualization distinguishes flow structures, by introducing seeders into local regions of the flow that are of interest to the study; this is *inhomogeneous* seeding (Westerweel, 1997). The local differences of particle concentration make it possible to distinguish the flow that is carrying the seeders from the surrounding fluid. For example, a turbulent jet that is carrying the seeders interacts with the surrounding fluid where there are no seeders present. Inhomogeneous seeding would therefore allow the flow structures forming at the interface between the jet and the surrounding fluid to become visible.

An important aspect of implementing planar laser imaging is the type of substance or material that is used to seed the flow, for this affects the illuminating mechanism of the seeding particles, of which there are two main branches:

1. Planar Laser-Induced Fluorescence (PLIF).
2. Direct scattering.

The PLIF method involves using a certain type of seeding material which absorbs the laser energy; in response the molecules of this material are excited and *fluoresce* (Smits and Lim, 2000). The intensity of the fluorescence is significant enough that it may be captured on camera. Therefore, the seeders that are excited by the laser illumination are now visible in the recordings. Hence, the flow structure is revealed. PLIF methods are typically well suited to water tunnel experiments, where a fluorescent dye such as Rhodamine 6G (99%) is injected into the flow at a location important to the investigation. When the dye is excited by the laser, the flow structure is clearly identifiable, where it typically has a metallic green appearance (Wang and Zhou, 2005).

The direct scattering illumination mechanism is caused when the high-energy laser light interacts with the seeder particles, which then result in laser light scattering in the direction of the CCD camera. The intensity of the scattered light should be sufficient to cause high enough exposure of the CCD detector to make the particles clearly visible in the recordings. The level of exposure is mainly related to (1) the particle diameter, (2) refractive index of the particles, (3) the energy of the laser light source, and (4) concentration of the seeders in the flow (Adrian and Westerweel, 2011). The behavior of the scattered light is described by Mie Scattering Theory, which reveals that the intensity of the scattered light increases strongly with particle diameter. However, this need must be balanced with the need of the particles to closely follow the fluid flow path, for which smaller particles that match the density of the working fluid are desirable (ie, neutrally buoyant particles). Therefore careful consideration of the seeder material is required.

Example 2.4: Multiple Jets Impinging on Side-by-Side Cylinders

A. Background

The flow structure of a submerged, fully laminar free jet defined within the range $300 < Re_{Dj} < 1000$ has been shown to travel great distances away from the nozzle exit without interacting with the surrounding fluid. These types of jets typically have parallel sides where there is no apparent mixing or mass transfer occurring between the jet and the surrounding fluid (McNaughton and Sinclair, 1966). As the Reynolds number is increased beyond the laminar regime, the distance at which the laminar jet remains stable decreases and it may be observed that certain eddy structures develop; this is the semiturbulent jet behavior where $1000 < Re_{Dj} < 3000$. Previous studies have mainly characterized the fully laminar free jet behavior. However, there are relatively few works considering how this behavior is altered when a target surface is placed in the path of the laminar jet, such as a circular cylinder. In addition, if multiple jets and cylinders are arranged in rows, where the jets are a certain distance ahead of the cylinders, the effects of proximity of the cylinders and jets on the laminar behavior are not well understood. To this end, flow visualization experiments have been conducted to understand (1) the flow structure of a row of fully laminar jets interacting with stationary surrounding fluid upstream of a row of cylinders, and (2) how this fully laminar jet behavior is altered when the Reynolds number is increased.

B. Experimental Details

The planar laser imaging using the direct scattering method was applied to the multiple impinging jet configuration shown in Fig. 2.26a illustrates a plan view of the experimental setup where the center axis of the round jet and the plane of the laser light disk is perpendicular to the center axis of the cylinders. Fig. 2.26b shows five round jets of diameter $D_j = 6.0$ mm impinged on a row of five side-by-side circular cylinders, where the jets' central axis was coincident with the central axis of the cylinders and the

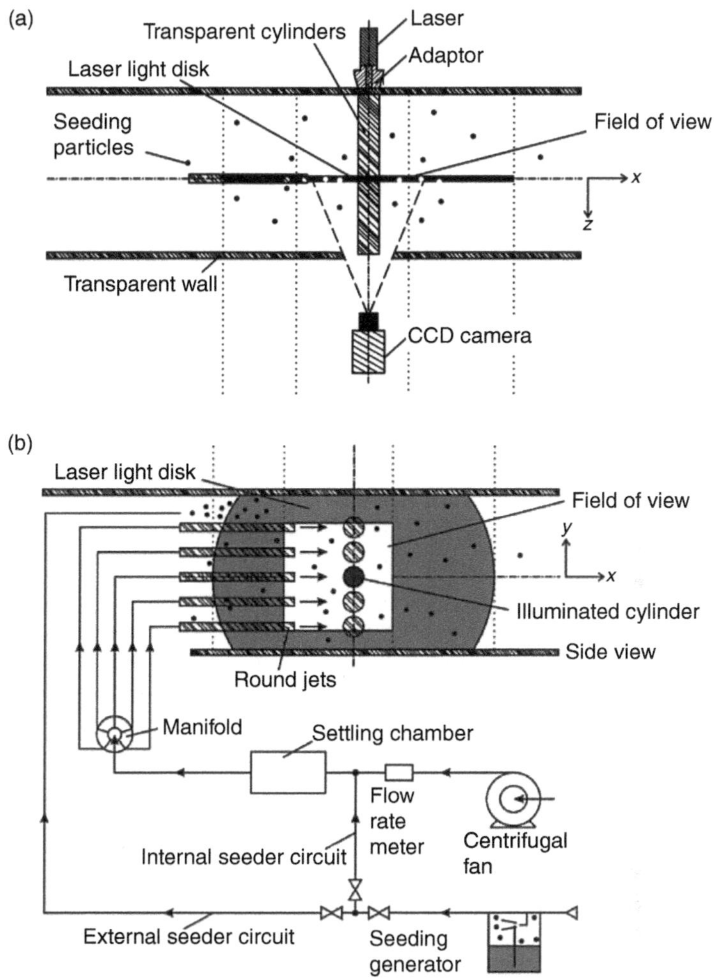

Figure 2.26 ***Schematic of the test setup for flow visualization of a laminar jet row interacting with a row of circular cylinders.*** (a) Plan-view of the experimental setup. (b) Side-view of the experimental setup detailing the system used for feeding seeder particles into the test rig.

spacing ratio was $T/D = 1.7$ where T is the transverse center-to-center spacing between the cylinders and D is the cylinder diameter equal to 15.0 mm. The jet Reynolds number was based on the flow rate feeding air to the jets and was set to $Re_{Dj} = 730$ and $Re_{Dj} = 2780$, corresponding to the fully laminar and semiturbulent jet flow regimes, respectively (McNaughton and Sinclair, 1966). The air was supplied to the jet nozzles with a centrifugal fan via a circular manifold with five outlets (ie, 72° spacing), where the tubes coming off the manifold feeding the nozzles had equal lengths. The jets were spaced $6D_j$ ahead of the cylinders.

The region where the jet interacts with the surrounding fluid and the cylinders is illuminated using the internal illumination method, where the planar laser light source originates from the center of the middle cylinder as shown in Fig. 2.26b. The direct scattering method was used to visualize the flow field using seeding particles atomized from mineral oil using a seeding generator, with a mean particle diameter of $d_p \approx 1.0$ μm.

The light sheet was adjusted to be extremely thin, $\Delta z \approx 0.5$ mm, so the jet flow features could be made more visible. A thicker light sheet means that the out-of-plane motion of the particles will reduce the resolution of the flow structures captured in the recording. More details on the internal illumination configuration are provided in Chapter 5. The field of view was set (ie, 110 mm × 110 mm) by positioning the camera a certain distance from the plane of the laser sheet and adjusting the camera lens' focus.

In the schematic two methods of introducing the seeders into the test rig are illustrated: (1) the internal seeder circuit and (2) the external seeder circuit. As has been previously stated, flow visualization relies on inhomogeneous seeding to reveal flow features. This may be achieved by firstly introducing the seeder particles into the air supply of the jets (ie, internal seeder circuit). However, the addition of momentum into the jet supply from the seeding generator would alter the Reynolds number and make the flow extremely unsteady. In order to prevent a high momentum injection of the seeders, which may also disturb the jet behavior, a high-volume settling chamber was filled with the seeding particles prior to the acquisition of the flow images. During the recordings the seeding generator was isolated using a shut-off valve.

The external seeder circuit was much simpler to implement, and involved seeding the quiescent surrounding air with seeding particles, while the jet air supply was clean.

C. Discussion of Results

Fig. 2.27a reveals the flow structure of a laminar jet interacting with the surrounding fluid. In this case the external seeder circuit was used and inhomogeneous seeding is observed, where the surrounding fluid has the seeding particles that scatter the laser illumination originating from the middle cylinder whereas the jet flow contains no seeders. The fully laminar jet has parallel edges, consistent with the description of McNaughton and Sinclair (1966). There is no apparent entrainment of the fluid prior to the cylinders. However, on the aft surface of the cylinders, the laminar boundary layer formed around the cylinders appears to break down and eddy flow structures develop. Consequently the surrounding fluid is entrained by the jet flow only on the aft surface of the cylinder as a result of the formation of eddy flow structures. In acquiring these flow images, the laser power can be adjusted to find the ideal exposure level from the scattered light. Overly high laser power potentially causes blooming of the CCD camera and this should be avoided, as it may damage the CCD camera.

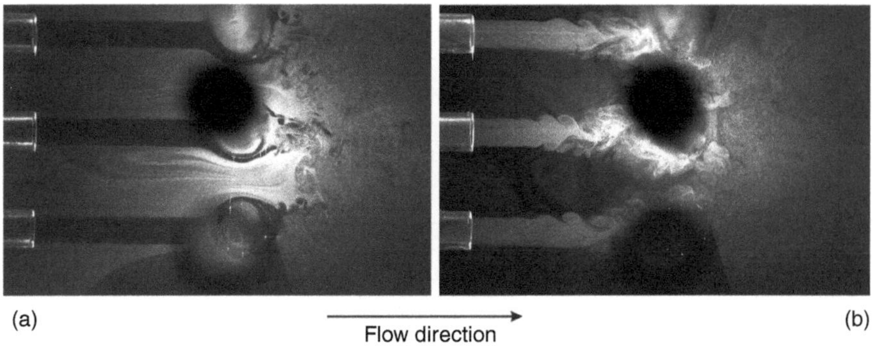

Figure 2.27 ***Flow visualization images of a row of five round jets impinging a row of five side-by-side circular cylinders.*** (a) Fully laminar jet, $Re_{Dj} = 730$, using the external seeding circuit; (b) semiturbulent jet, $Re_{Dj} = 2780$, employing the internal seeding circuit.

Fig. 2.27b shows the flow image generated by introducing the seeding particles with the jet flow (ie, internal seeder circuit). In this instance the higher Reynolds number leads to the semiturbulent jet. In contrast to the fully laminar jet (Fig. 2.27b), eddy structures now develop ahead of the cylinders, indicating that the laminar jet is no longer stable when the fluid momentum has been increased. Initially, the jet develops sinusoidal undulations from $\approx 1D_j$ from the jet nozzle exit, which appear to increase in amplitude until eddy structures eventually form at $\approx 2.5D_j$. As the jet progresses further downstream, the jet transitions to fully turbulent behavior ahead of the cylinders when $\approx 5D_j$, where there is substantial flow mixing and diffusion of the jet. It may be apparent from the images that although the jet structure is visible the quality of the images may be improved significantly, by ensuring that the surrounding fluid does not contain the seeding particles that can also scatter the laser light towards the camera. This may be achieved by taking images earlier in the sequence, before the concentration of the seeding particles increases to the point where they also scatter light. As the concentration of the seeding particles in the surrounding fluid is increased, there will eventually be a homogeneous distribution of seeders in the test section, making it increasingly difficult to distinguish the jet flow from the surrounding fluid.

2.7 IN-FLOW VISUALIZATION USING INK-DYE PIGMENT INJECTION

2.7.1 Background

Water flow around an object whose field is optically accessible can be visualized by an ink-dye pigment injection technique. This technique is inexpensive and easy to implement. The flow pattern is realized by injecting an ink neutrally buoyant with water flow. A classic example is the experiment of Osborne Reynolds, which showed for the first time the transition from laminar flow to turbulent flow in a water pipe. An ink-dye pigment with obvious contrast to the water background is preferable. However, most of the ink-dye pigment is readily dissipated when applied to a highly turbulent flow.

2.7.2 Specific Gravity of Ink-pigment

The selected ink-dye pigment should follow the flow with minimal disturbance to the flow as a result of the specific gravity (buoyancy force) match between the host medium (water) and the ink-dye pigment. Prior to its application, the specific gravity of the ink-dye pigment should be precisely measured.

2.7.3 Injection Velocity

When the ink-dye pigment is injected, there is a mismatch of injection velocity between the water and the pigment. As observed by Choi et al. (2007), when the exit velocity is too high, a jet flow is produced which generates "mushroom-like" structures. Similarly, when the exit velocity is too low, wake structures are formed, which appear as a series of interconnecting vortex loops. When a correct exit velocity is reached, the dye should appear as a smooth filament.

2.7.4 Injection Probe

Typically a hypodermic needle or steel tubing with small inner/outer diameters ranging from 0.5 to 2.0 mm can be used to inject ink-dye pigment at the proper velocity. The probe (needle or steel tubing) is then connected to a reservoir of ink-dye pigment by a flexible polycarbonate tube. The ink-dye pigment is released by gravity or external pressure. Due to the presence of the probe, the flow field to be visualized may be disturbed. To minimize the flow disturbance, the probe is often located some distance upstream from the flow area of interest. However, the streaklines formed by the injected oil-dye pigment are readily dissipated, depending on the flow pattern. A proper distance can be chosen after trial and error for a given flow velocity.

Another probe type is a probe that is fabricated as part of the test rig. This approach could eliminate the flow disturbance caused by the presence of the probe. However, only flow that departs from the test model as a result of the flow pattern around the test model, can be visualized. If flow around the test model is of interest, this probe type would not yield successful results.

2.7.5 Photographing

Recording a series of flow images could provide an image of an interesting flow pattern; simply selecting one from a series of digitized images could be compared to using snapshot images from a still camera. This is due to the fact that the interesting flow feature could come at any moment.

2.7.6 ETC

The selection of color for the oil-dye pigment should maximize the contrast with the background. For example, when a blue oil-dye pigment is chosen, placing a white board behind the water tunnel would enhance the contrast and hence provide a clearer flow pattern for visualization.

Example 2.5: Flow Around an Inclined Single Cylinder

A. Background

It is common to observe alternating eddies developing behind a bluff body, such as the Kármán eddy street, when a fluid flows past a stationary cylinder placed normal to the mainstream flow. Its development is strongly dependent on the Reynolds number. Existing literature regarding such phenomena is overwhelming (Zdravkovich, 1997). The pattern of the alternating eddies is expected to vary when the cylinder is inclined along the mainstream flow. This is attributed to the inclination-induced velocity component along the cylinder span if the cylinder span is relatively short, implying the invalidity of the "principle of independence" (Choi et al., 2007). In the present example, alternating eddy patterns resulting from the inclination of a short cylinder with both ends attached to endwall plates are qualitatively considered using the oil-dye injection technique.

B. Test Setup

B1. Test Rig

As a set of discs rotates, water is drawn by the friction between the disc surfaces and water. This scheme minimizes the upstream propagation of flow disturbances. For the disc assembly, a series of thin (3 mm thick) discs with 400 mm diameter is stacked with 3 mm gaps between them. Water flows through an open flow channel, guide vanes, and a contraction section. There are a honeycomb and a screen placed 300 mm upstream of the test section.

B2. Ink-dye Pigment Injection Setup

To observe flow around a cylinder of diameter D in a rectangular water tunnel made of Perspex plates, an ink-dye solution with a specific gravity similar to water is injected through a small steel tube (0.5 mm inner diameter), as illustrated in Fig. 2.28. The tube exit is located at $5D$ distance upstream from the cylinder. The streaklines are then captured through the transparent endwalls and sidewalls of the water tunnel by a digital camcorder.

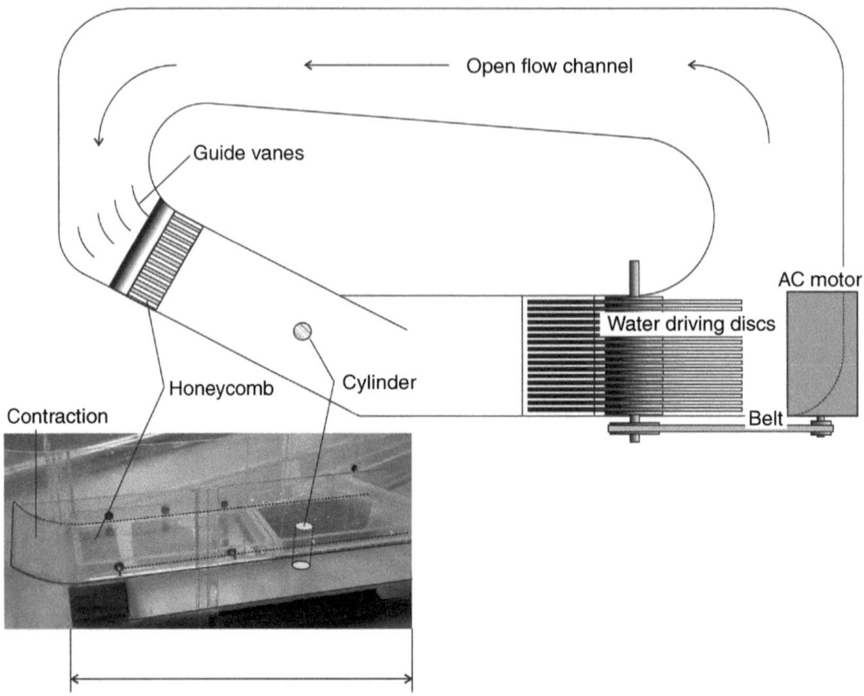

Figure 2.28 ***Water tunnel test facility with a set of frictional discs driving water current.***

C. Discussion of Results

Fig. 2.29a shows an instantaneous image of the midspan periodic wakes shed from an upright cylinder at $\mathrm{Re}_D = 1.0 \times 10^4$. The alternating eddies are clearly visible. Prior to the flow visualization, the wake pattern is quantitatively characterized by a dimensionless wake shedding frequency, that is, the Strouhal number ($S_t = fD/U$), where f is the shedding frequency measured at 2D downstream from the edge of the cylinder. In the present study, S_t was found to be 0.195 at $\mathrm{Re}_D = 1.0 \times 10^4$ for the noninclined (upright) cylinder, which agrees well with the data reported in the open literature (Zdravkovich, 1997).

Fig. 2.29b shows an instantaneous image of the 45° inclined cylinder where the wakes are shed from the midspan of the cylinder. Although it is not as clear as in the case of an upright cylinder (Fig. 2.29a), a periodic wake shedding from the cylinder can still be observed. On the other hand, the width between the alternating vortices increases in comparison with that of the noninclined cylinder.

Fig. 2.30 shows flow visualization results obtained from the water tunnel experiments. The trace of ink particles released upstream of the cylinder shows that the mainstream flow is skewed towards the upper endwall as it approaches

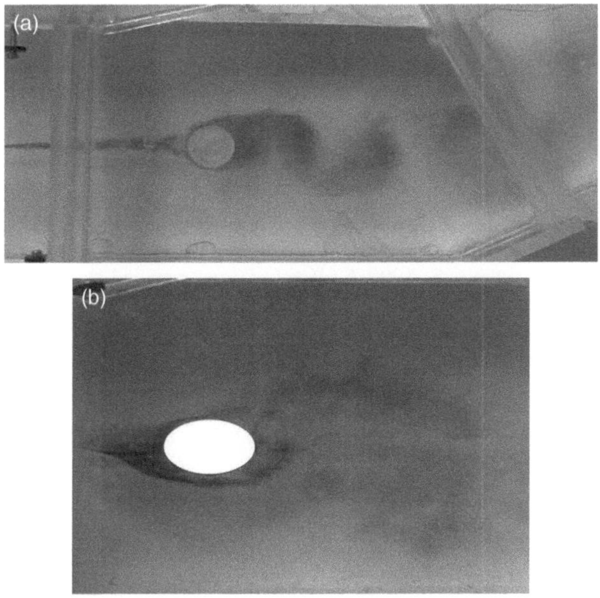

Figure 2.29 ***Instantaneous flow images of flow past single cylinder visualized by ink dye injection in water.*** (a) 0° inclined case; (b) 45° inclined case.

the leading edge of the cylinder. Some of the injected ink particles follow the contour of the cylinder before they are separated from the cylinder surface. The wakes shed from the cylinder feed the fluid with a spanwise velocity into the aft cylinder, forming a jet-like flow. This fluid is then driven towards the upper endwall and impinges onto it. This jet-like flow is the main source of the high-momentum flow region at the aft cylinder. Furthermore, since the jet-like flow behaves as a crossflow to the wakes shed from the cylinder, it induces increased mixing of the mainstream flow behind the cylinder.

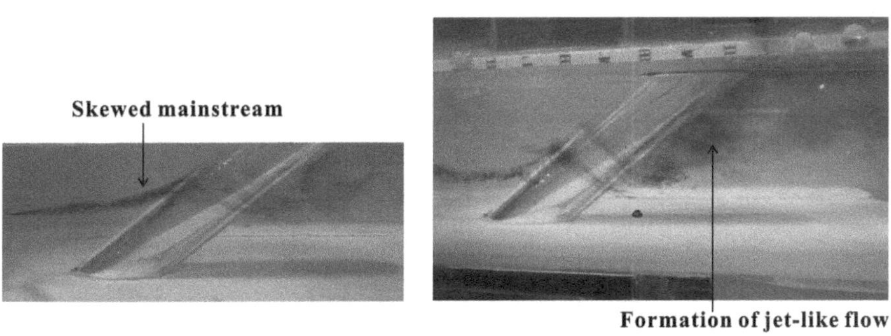

Figure 2.30 ***Flow patterns around 45° inclined cylinder.***

REFERENCES

Abramovich, G.N., 1963. The Theory of Turbulent Jets. MIT Press, Cambridge, MA, USA, 1-5.

Adrian, R.J., Westerweel, J., 2011. Particle Image Velocimetry. Cambridge University Press, Cambridge, UK.

ASTM E1862-97(2002)e1, 1997. Standard Test Methods for Measuring and Compensating for Reflected Temperature Using Infrared Imaging Radiometers. ASTM International, West Conshohocken, PA.

Baker, C.J., 1979. The laminar horseshoe vortex. J. Fluid Mech. 95, 347–367.

Baker, C.J., 1980. The turbulent horseshoe vortex. J. Wind Eng. Ind. Aerod. 6, 9–23.

Ballio, F., Bettoni, C., Franzetti, S., 1998. A survey of time-averaged characteristics of laminar and turbulent horseshoe vortices. J Fluid Eng. – T. ASME 120, 233–242.

Baughn, J.W., Anderson, M.R., Mayhew, J.E., Wolf, J.D., 1999. Hysteresis of thermochromic liquid crystal temperature measurement based on hue. J. Heat Transf. 121, 1067–1072.

Camci, C., Kim, K., Hippensteele, S.A., 1992. A new hue capturing technique for the quantitative interpretation of liquid crystal images used in convective heat transfer studies. J. Turbomach. 114, 765–775.

Choi, I.K., Kim, T., Song, S.J., Lu, T.J., 2007. Endwall heat transfer and fluid-flow around an inclined short cylinder. Int. J. Heat Mass Tran. 50, 919–930.

Gardon, R., Akfirat, J.C., 1965. The role of turbulence in determining the heat-transfer characteristics of impinging jets. Int. J. Heat Mass Tran. 8, 1261–1272.

Gauntner, J.W., Livingood, J.N.B., Hrycak, P., 1970. Survey of Literature on Flow Characteristics of a Single Turbulent Jet Impinging on a Flat Plate, NASA TN D-5652, NTIS N70-18963. NASA, Lewis Research Center, Washington, DC.

Giralt, F., Chia, D.-J., Trass, O., 1977. Characterization of impingement region in an axisymmetric turbulent jet. Ind. Eng. Chem. Fund. 16, 21–28.

Hofmann, H.M., Kind, M., Martin, H., 2007. Measurements on steady state heat transfer and flow structure and new correlations for heat and mass transfer in submerged impinging jets. Int. J. Heat Mass Tran. 50, 3957–3965.

Hung, C.M., Sung, C.H., Chen, C.L., 1991. Computation of Saddle Point of Attachment, AIAA Paper 91-1713.

Hunt, J.C.R., Abell, C.J., Petetka, J.A., Woo, H., 1978. Kinematical studies of the flows around free or surface-mounted obstacles; applying topology to flow visualization. J. Fluid Mech. 86, 179–200.

Ireland, P.T., Wang, Z., Jones, T.V., 1995. Measurement Techniques: Liquid Crystal Heat Transfer Measurements, von Karman Institute for Fluid Dynamics, Lecture Series 1995-01.

Ireland, P.T., Neely, A.J., Gillespie, D.R.H., Robertson, A.J., 1999. Turbulent heat transfer measurement using liquid crystals. Int. J. Heat Fluid Fl. 20, 355–367.

Jambunathan, K., Lai, E., Moss, M.A., Button, B.L., 1992. A review of heat transfer data for single circular jet impingement. Int. J. Heat Fluid Fl. 13, 106–115.

Khan, M.J., Ahmed, A., 2005. Topological model of flow regimes in the plane of symmetry of a surface-mounted obstacle. Phys. Fluids 17, 045101.

Khan, M.J., Ahmed, A., Trosper, J.R., 1995. Dynamics of the juncture vortex. AIAA J. 33, 1273–1278.

Kim, T., 2004. Fluid-flow and Heat-transfer in a Lattice-Frame Material, PhD Thesis, Department of Engineering, University of Cambridge.

Kim, T., Hodson, H.P., Lu, T.J., 2004. Fluid-flow and endwall heat-transfer characteristics of an ultralight lattice-frame material. Int. J. Heat Mass Tran. 47, 1129–1140.

Martin, H., 1977. Heat and mass transfer between impinging gas jets and solid surfaces. Advances in Heat Transfer 13, 1–60.

McNaughton, K.J., Sinclair, C.G., 1966. Submerged jets in short cylindrical flow vessels. J. Fluid Mech. 25 (2), 367–375.

Merchant, J., 2012. Infrared Temperature Measurement Theory and Application. Mikron Instrument Company Inc.
Raffel, M., Willert, C.E., Kompenhans, J., 1998. Particle Image Velocimetry: A Practical Guide. Springer, Heidelberg.
Reungoat, D., Riviere, N., Faure, J.P., 2007. 3C PIV and PLIF measurement in turbulent mixing: Round jet impingement. Visual. – Japan 10, 99–110.
SAI, 2014. Helium Bubble Generator Systems for Airflow Visualization Manual. Sage Action Inc., Apache Junction, AZ, 6.
Simpson, R.L., 2001. Junction flows. Annu. Rev. Fluid Mech. 33, 415–443.
Smits, A.J., Lim, T.T., 2000. Flow Visualization: Techniques and Examples. Imperial College Press, London.
Tobak, M., Peake, D.J., 1979. Topology of two-dimensional and three-dimensional separated flows, AIAA 12th Fluid and Plasma Dynamics Conference, AIAA paper 79-1480.
Usamentiaga, R., Venegas, P., Guerediaga, J., Vega, L., Molleda, J., Bulnes, F.G., 2014. Infrared thermography for temperature measurement and non-destructive testing. Sensors – Basel 14, 12305–12348.
Visbal, M.R., 1991. Structure of laminar juncture flows. AIAA J. 29, 1273–1282.
Viskanta, R., 1993. Heat transfer to impinging isothermal gas and flame jets. Exp. Therm. Fluid Sci. 6, 111–134.
Vollmer, M., Mollmann, K.P., 2010. Infrared Thermal Imaging: Fundamentals, Research and Applications. Wiley, Hoboken, NJ.
Wang, Z.J., Zhou, Y., 2005. Vortex interactions in a two side-by-side cylinder near wake. Int. J. Heat Fluid Fl. 26, 362–377.
Wang, Z., Ireland, P.T., Jones, T.V., 1995. An advanced method of processing liquid crystal video signals from transient heat transfer experiments. J. Heat Transf. 117, 184–189.
Wang, Z., Ireland, P.T., Jones, T.V., Davenport, R., 1996. A color image processing system for transient liquid crystal heat transfer experiments. J. Turbomach. 118, 421–427.
Wang, Z., Ireland, P.T., Kohler, S.T., Chew, J.W., 1998. Heat transfer measurements to a gas turbine cooling passage with inclined ribs. J. Turbomach. 120, 63–69.
Westerweel, J., 1997. Fundamentals of digital particle image velocimetry. Meas. Sci. Technol. 8, 1379–1392.
Yang, G., Choi, M., Lee, J.S., 1999. An experimental study of slot jet impingement cooling on concave surface: Effects of nozzle configuration and curvature. Int. Heat Mass Tran. 42, 2199–2209.
Zdravkovich, M.M., 1997. Flow Around Circular Cylinders, Vol. 1: Applications. Oxford Science Publications, Oxford.

CHAPTER 3

Pneumatic Measurements for Pressure, Velocity, and Flow-direction

K. Kim
School of Electrical and Information Engineering, University of the Witwatersrand, Johannesburg, South Africa

3.1 INTRODUCTION

Pressure measurements using pneumatic tappings (probes) are reliable compared to other means. The pressure data from the pneumatic probes are considered to be "steady-state" due to their relatively slow response to the changes in flow conditions. When a pressure transducer records variation of fluctuation, the pressure data tend to be averaged out in time.

In many cases, the pressure values relative to a reference pressure rather than absolute pressure values are of concern, such as pressure drop through a heat exchanger element. Furthermore, based upon the pressure data, flow velocity and direction of flow can also be calculated. In this chapter, the following four categories are considered separately:

1. Static pressure
2. Stagnation (total) pressure
3. Flow velocity
4. Flow direction

based on pneumatically measured static and stagnation (or total) pressure data.

3.2 STATIC PRESSURE MEASUREMENT

At boundaries between a solid and a fluid, a pressure-measuring device is often termed "pressure tapping." The static pressure tappings are small holes drilled into the flow boundary material (eg, solid wall). Typically, a circular tube connects the pressure hole, which is perpendicular to the boundary wall, and a pressure transducer. The diameter of the pressure tapping (hole) should be significantly smaller than the momentum thickness of the boundary layer above the tapping (Rayle, 1959). Some of the inaccuracies associated with

Application of Thermo-Fluidic Measurement Techniques
http://dx.doi.org/10.1016/B978-0-12-809731-1.00003-4

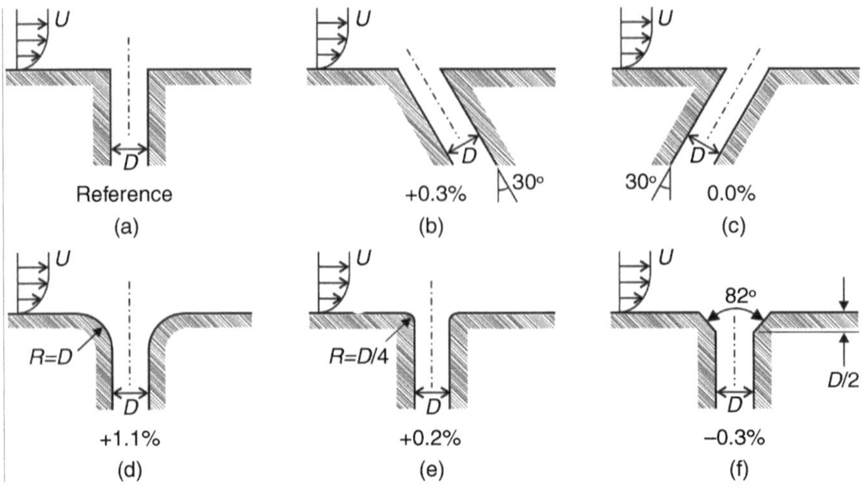

Figure 3.1 ***Configuration of static pressure tapping and effect of its geometry on measurement accuracy (Rayle, 1959).***

different tapping configurations are illustrated in Fig. 3.1a–f. The error in the figure is relative to the dynamic pressure measured by a sharp-edged tapping with the same diameter (Fig. 3.1a).

In heat exchanger applications where a heat exchanger core (eg, fins and metal foam) is inserted in a rectangular or tubular flow channel, the mainstream flow is internal to the channel but external to the inserted media. In such cases, how great a pressure drop through the heat exchanger core occurs is important, in addition to overall heat transfer. To this end, the pressure tappings are drilled into the channel wall at the inlet and outlet sections. In addition, other tappings are often inserted along the mainstream direction to obtain the longitudinal variation of the static pressure.

Fig. 3.2a–c depict three selected cases of the mainstream flow through different channel flow configurations. Consider first, flow through an empty channel with "constant flow area" as shown in Fig. 3.2a. The downstream pressure (p_2) is lower than the upstream pressure (p_1) due to pressure drop (loss) from skin friction between the mainstream and the channel wall. However, the pressure drop is typically negligible if the channel length L is short relative to the channel height H. Flow velocity at the inlet is almost the same as that at the outlet if displacement thickness, which acts to reduce the effective flow area of the channel, is thin in comparison with H (Schlichting, 1979).

For the case when the flow area is linearly decreasing as shown in Fig. 3.2c, the pressure difference $\Delta p = p_1 - p_2$ becomes obvious. Again, we assume negligible displacement thickness and pressure loss due to skin friction. The

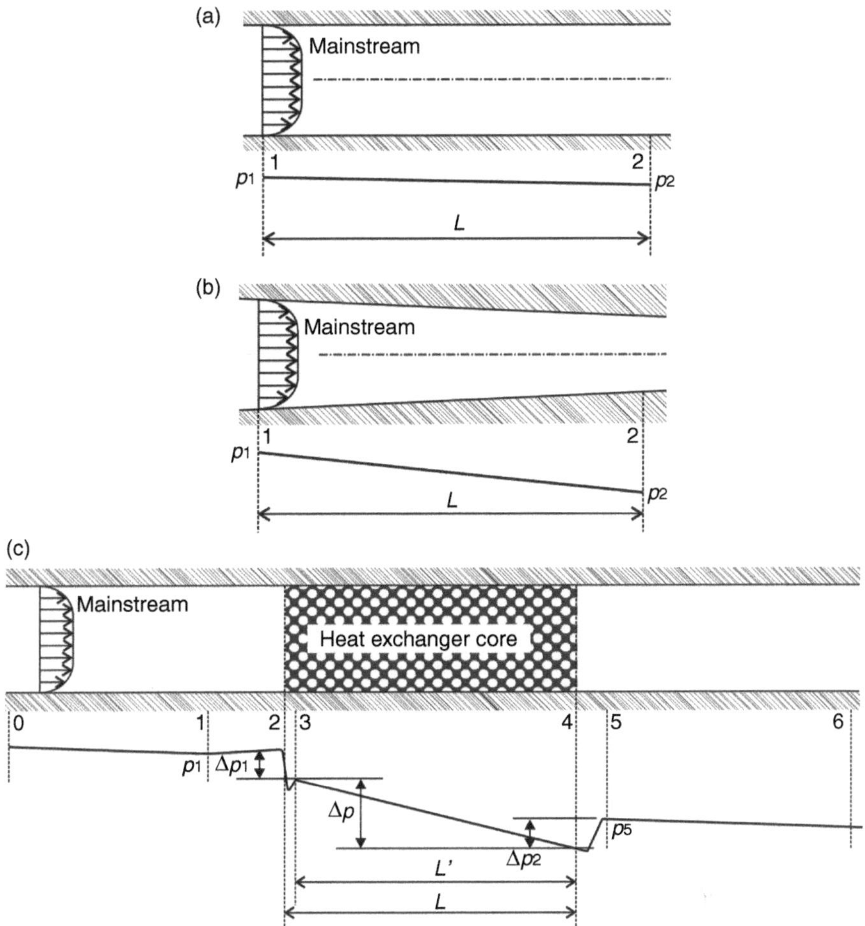

Figure 3.2 ***Pressure drop through various flow channel configurations.*** (a) Constant flow area; (b) flow area contraction; (c) constant flow area channel with a heat exchanger core insertion (Kays and London, 1984).

pressure difference (or pressure drop) still exists, resulting from the reduced flow area along the mainstream direction, not from pressure loss (ie, an irreversible process). In this case, the amount of decreased static pressure is converted into dynamic pressure (or flow velocity), so that the total pressure (static pressure + dynamic pressure) is conserved.

Finally, consider flow through a constant flow area channel containing a heat exchanger core element as illustrated in Fig. 3.2c. Before the mainstream reaches the heat exchanger core, the static pressure slightly decreases due to skin friction ($0 \rightarrow 1$) but increases before the flow enters the heat exchanger core (1 → 2) since the mainstream stagnates on the core front. Because the inserted core element reduces the effective total flow area in this region, the static pressure drops, Δp_1 in (2 → 3). However, this pressure drop is not only caused by flow area change due to core insertion alone, but also affected by the abrupt contraction that induces boundary layer separation. This phenomenon is collectively called the "entry effect."

As illustrated in Fig. 3.2c, as the mainstream flows across the heat exchanger core, the static pressure decreases linearly till the exit of the core (ie, $\Delta p/L'$ = constant), followed by a sudden increase in the static pressure. This increased static pressure is a consequence of two contributors: (1) the increased flow area increases the static pressure and reduces the flow velocity accordingly, and (2) irreversible expansion (boundary layer separation) to the surroundings decreases the static pressure. This is collectively called the "exit effect."

The variation of static pressure along the direction of the mainstream in the three cases demonstrates that the pressure readings from each static pressure tapping differ, depending on the longitudinal location chosen. This indicates the importance of the proper selection of pressure tapping locations.

Example 3.1: Flow Separation From a Circular Cylinder

A. Background

Flow around a circular cylinder is a classical and longstanding subject. The flow around an infinitely long cylinder with a circular cross-section can be treated as 2D (ie the velocity component parallel to the cylinder axis plays no part in the flow; Zdravkovich, 1997, 2003; Sparrow and Yanez Moreno, 1987). The flow external to the cylinder approaches the cylinder in crossflow (relative to the cylinder axis).

The mainstream flow stagnates on the leading edge of the cylinder, then follows the cylinder contour before it is separated from its surface if the Reynolds number is high enough to cause separation. The azimuth angle of the separation, if it occurs, varies with the Reynolds number. It is also dependent upon the pressure gradient along the circumference of the cylinder. The example in Fig. 3.3a and b demonstrates the effects of the stream-wise pressure gradient on flow separation from a single circular cylinder. To impose the stream-wise pressure gradient, the method of sidewall contouring (Fig. 3.3b) is used for both an adverse pressure gradient and a favorable pressure gradient.

B. Test Setup

For circumferential pressure distribution on the cylinder, a single static pressure tapping with inner diameter of 0.5 mm was drilled into the cylinder

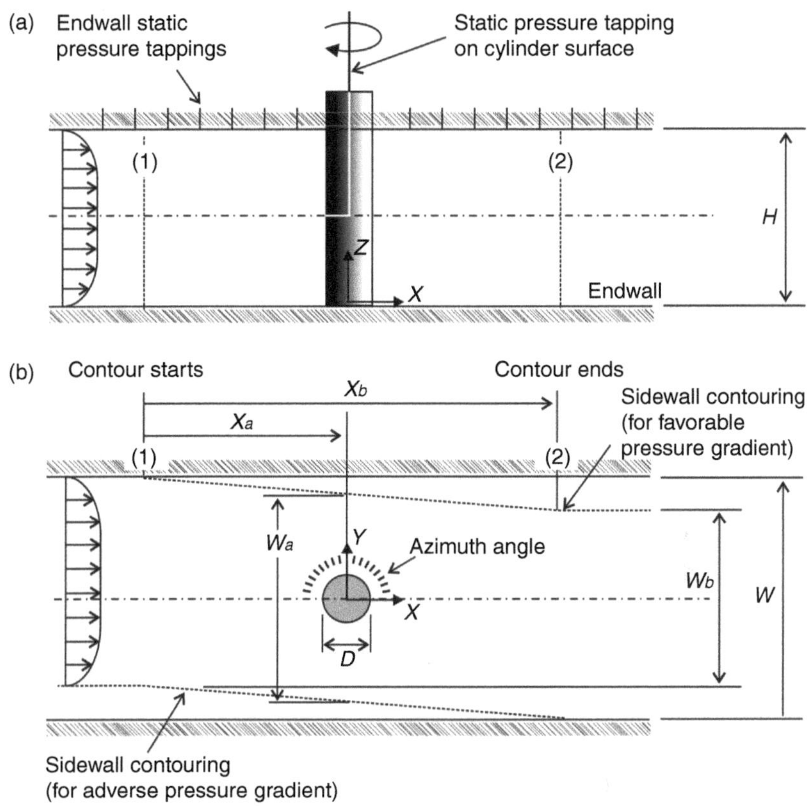

Figure 3.3 ***Schematic of experimental setup for a single static pressure tapping in a cylinder connected with a rotary table under a stream-wise pressure gradient by sidewall contouring.*** (a) Side view; (b) plan view.

(made of Perspex) midspan. The cylinder axis was connected to a rotary table allowing 360° rotation. During the experiment, the rotation was manually performed with 5° increments. The pressure tapping was connected to a differential pressure transducer (DSA™, Scanivalve Inc.). Detailed parameters of the present experimental setup are listed in Table 3.1.

To evaluate the distribution of circumferential pressure on the cylinder, a nondimensional pressure coefficient (C_p) is defined as:

$$C_p = \frac{p(0) - p(\varphi)}{\rho U_{Xa}^2/2} \tag{3.1}$$

where $p(0)$ and $p(\varphi)$ are the static pressures measured along the cylinder circumference at $\varphi = 0°$ and an arbitrary azimuth angle (φ), and U_{Xa} is the calculated flow velocity at $x = X_a$ using Bernoulli's equation based

Table 3.1 Parameters of the Cylinder and Flow Channel Considered

Cylinder	Diameter, $D = 0.022$ m Span, $l = 0.12$ m	Aspect ratio, $l/D = 5.45$
Channel	Width at station (1), $W = 0.14$ m Width at station (2), $W_b = 0.112$ m Width at $x = X_a$, $W_a = 0.126$ m Height, $H = 0.12$ m	Blockage ratio at $x = X_a$ $W_a/D = 5.73$ $X_a/D = 5.0$ $X_b/D = 10.0$

on the flow area change and total mass flow rate. The origin of the azimuth angle (φ) is coincident with the stagnation point. As reference, the pressure coefficient in potential (frictionless) flow is defined (Katz and Plotkin, 2004) as:

$$C_p = 1 - 4\sin^2(\varphi) \tag{3.2}$$

C. Discussion of Results

The circumferential variation of static pressure on the cylinder for three different cases is plotted in Fig. 3.4a together with that of potential flow (frictionless flow). In general, all of the cases appear to behave in a similar manner: the flow stagnates at zero azimuth angle, followed by a monotonic decrease up to about $\varphi = 65°$, and then increases, passing an inflection point ($\varphi \sim 75°$) where flow separation takes place (Zdravkovich, 1997). Slightly after the inflection point, the pressure is maximized before decreasing slowly to $\varphi = 180°$. Here, as the time-averaged flow is symmetric with respect to $\varphi = 180°$, only half of the circumference is covered (Zdravkovich, 1997, 2003).

No visible difference has been observed due to the imposed pressure gradient. The most sensitive flow feature is perhaps the flow separation point. To examine this, pressure distribution near the inflection point is highlighted in Fig. 3.4b. The shifting of the flow separation point is now clearly noticeable. Without the pressure gradient, flow separation takes place at $\varphi \sim 75°$. An adverse pressure gradient causes an earlier separation at $\varphi \sim 70°$ whereas a favorable pressure gradient delays the separation to $\varphi \sim 80°$.

In summary, the overall behavior of the circumferential pressure is almost insensitive to the stream-wise pressure gradient formed around the cylinder in the present experimental conditions. However, the separation of flow from the cylinder is delayed when a favorable pressure gradient is imposed and brought forward when an adverse pressure gradient is applied.

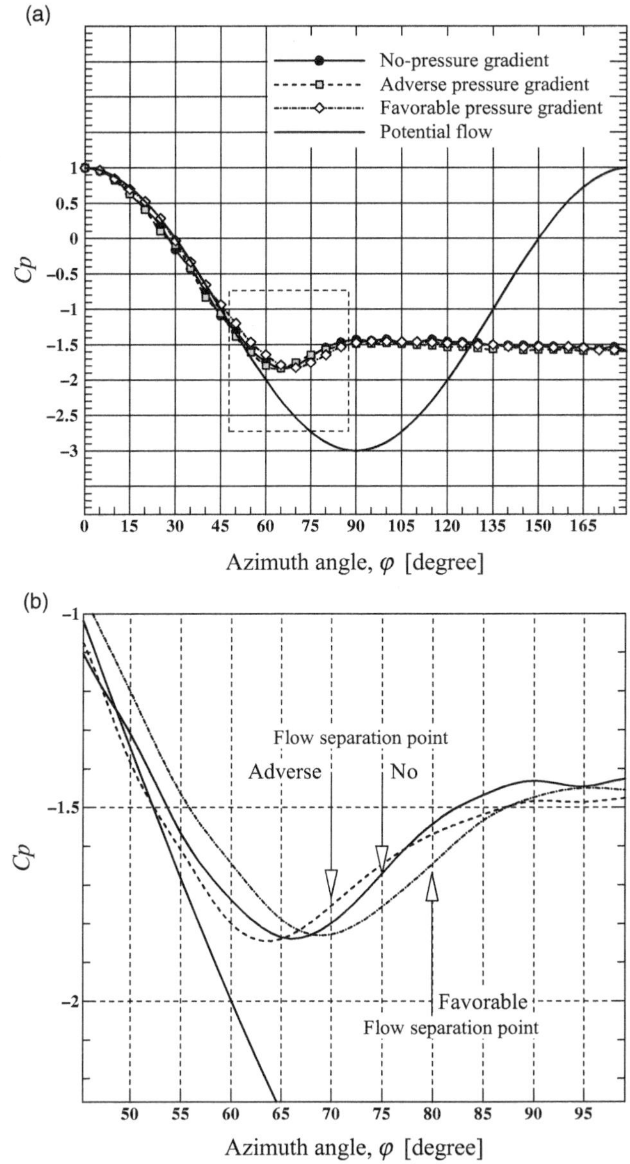

Figure 3.4 ***Circumferential pressure distributions for three selected cases; no pressure, adverse pressure, and favorable pressure gradients at a fixed mass flow rate of*** $\dot{m} = 0.14$ ***kg/s.*** (a) Overall variation; (b) separation point.

3.3 STAGNATION (TOTAL) PRESSURE MEASUREMENT

3.3.1 Background

Stagnation (or total) pressure is the pressure measured at a point in a fluid flow at which the fluid comes to rest and has the highest value found anywhere in the flow field. At the stagnation point, the velocity of the fluid flow is zero due to the fact that the kinetic energy is converted into pressure isentropically. Therefore, using Bernoulli's equation where viscous effects are neglected, the stagnation pressure (p_t) can be expressed as:

$$p_t = p_\infty + \frac{1}{2}\rho U_\infty^2 \tag{3.3}$$

where p_∞ is the static pressure and $\rho U_\infty^2/2$ is the dynamic pressure measured far upstream.

In common practice, stagnation pressure is measured intrusively by placing a half-open tube's end (or nose) in the direction of the oncoming flow. This will interfere with the flow near to the measurement point. Stagnation occurs instantaneously at the tube's nose. This allows most thin tubes of various designs, collectively known as pressure tubes or probes, to measure the stagnation pressure provided its axis is aligned with the known flow direction. Therefore, the measured pressure is equal to the local total pressure.

Three typical types of stagnation pressure probes are shown in Fig. 3.5a–c. A cylinder probe (Fig. 3.5a) can be constructed using a cylindrical metallic tube. The tube has a cylindrical nose with a diameter of, for example, 0.8 mm and is bent 90° so that the tube's opening is placed parallel to the flow. A Pitot probe (Fig. 3.5b) is also called a Pitot-static tube because it can measure both the static pressure and the stagnation pressure. The stagnation pressure is measured at the stagnation pressure port located at the nose of the probe whose axis is parallel to the nominal flow direction. The static pressure is measured at the static pressure port, which is located along the side of the Pitot probe where the port is oriented normal to the fluid stream. A Kiel probe (Fig. 3.5c) is similar to the cylinder probe in Fig. 3.5a, but it has a Venturi-like shroud attached to the inlet port. The Kiel probe allows stagnation pressure in flows of variable or unknown exact direction to be measured.

Since the cylinder type probe and the Kiel probe do not have a built-in static tapping unlike the Pitot probe, the static pressure has to be measured elsewhere if the velocity at a point must be measured. In this case, a static pressure tapping needs to be inserted flush with a representative surface,

(a)
(b)
(c)

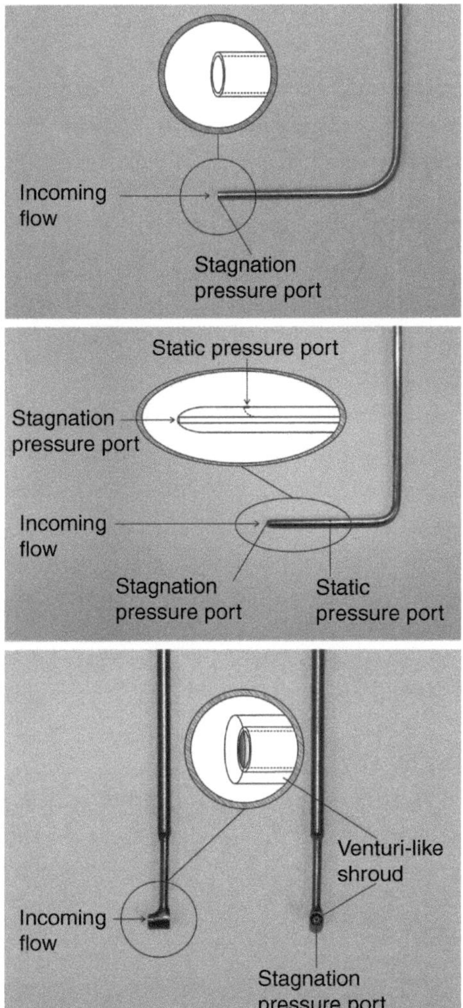

Figure 3.5 ***Some stagnation pressure probes.*** (a) Cylinder-type probe; (b) Pitot tube; (c) Kiel probe.

which is parallel to the nominal flow direction. The longitudinal location of the static tapping needs to be the same as that of the stagnation pressure port (details to be discussed in section 3.4).

Parameters that influence the measurements of stagnation pressures include:

1. incidence (or yaw) angle
2. Reynolds number

3. Mach number
4. velocity gradients
5. unsteadiness of the fluid flow

In the following sections, some aspects of these parameters are briefly discussed following Gracey (1957).

3.3.2 Yaw (Incidence) Angle Effect

In most aerodynamic measurements, the exact direction of fluid flow is essentially unknown although its nominal direction may be guessed. Misalignment of the stagnation pressure probe's axis (ie, not parallel to the flow direction) may cause errors. The degree of these errors is related to the type of the pressure probes. Therefore, prior knowledge of a given probe when performing the stagnation pressure measurements may minimize such errors.

To examine the effect of a yaw (or incident) angle on the stagnation pressure values, the three probes (in Fig. 3.5) are selected and calibrated for various yaw angles. Results are plotted in Fig. 3.6 showing the variation of the pressure coefficient C_p measured at the geometrical stagnation point. A nondimensional pressure coefficient (C_p) is defined as:

$$C_{p,yaw} = \frac{p_t(\varphi) - p_\infty(0)}{\rho U_\infty^2/2} \tag{3.4}$$

where $p_t(\alpha)$ is the stagnation pressure measured as the yaw angle is varied from $\alpha = 0°$, $p_\infty(\alpha)$ is the static pressures measured at the same stream-wise location as the stagnation tube, and U_∞ is the mean mainstream velocity far upstream.

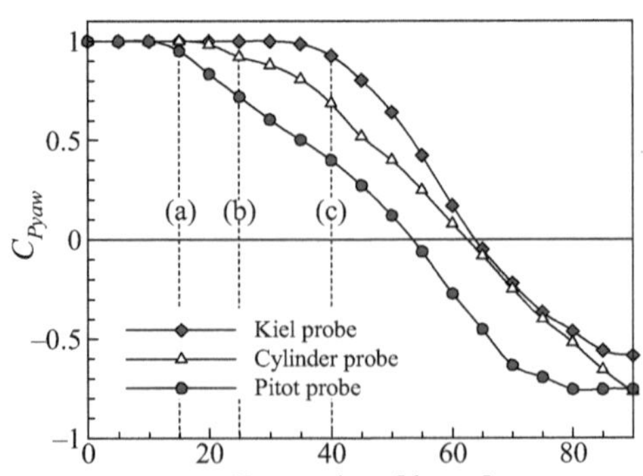

Figure 3.6 ***Incident (or yaw) angle ranges for three selected stagnation pressure probes.*** (a) Pitot probe, (b) cylinder probe, and (c) Kiel probe.

In general, the pressure is symmetric with respect to the reference line, that is, zero yaw lines. As the yaw angle is varied from $\alpha = 0°$, the pressure coefficient C_p from each pressure probe shows a different response. The Pitot probe shows very little tolerance to misalignment and can tolerate variations in the flow angle up to (a) $\alpha = 15°$. The cylinder-type probe can tolerate variations in the flow angle up to (b) $\alpha = 25°$ whilst the Kiel probe can tolerate up to (c) $\alpha = 40°$.

For the Kiel probe, the Venturi-like shroud that is attached around the inlet port of the Kiel probe forces the local flow to align itself with the shroud axis so that the flow is guided directly toward the measurement point. This provides a wider yaw angle coverage, up to $\alpha = 40°$ (for this selected model of the Kiel probe), than the other types of total pressure probes. Therefore, the Kiel probe is better suited for total pressure measurements in flows of variable or unknown exact direction.

3.3.3 Reynolds Number Effect

The stagnation pressure measurement is influenced by the Reynolds number because there is a viscous interaction between the free stream and the stagnating fluid. Therefore, an energy transfer due to momentum occurs between these two that results in a higher than normal stagnation pressure measurement. Thompson and Holder (1958) reported that the measurements are typically insensitive to a Reynolds number effect except at extremely low Reynolds numbers where a measurement with a Pitot tube of flow with $Re_d < 100$ is shown in Fig. 3.7. The pressure coefficient C_p was defined as:

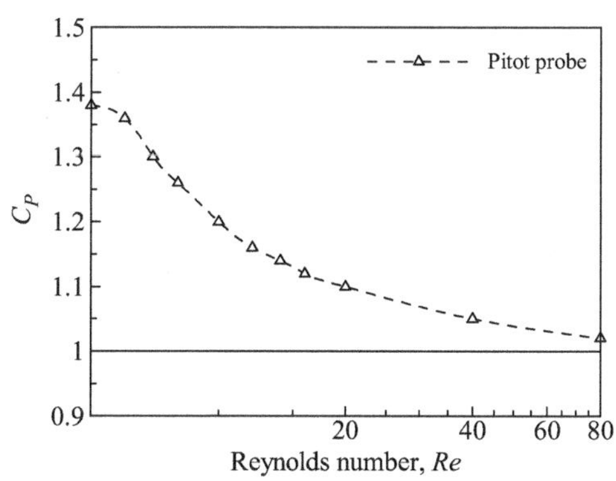

Figure 3.7 ***Calibration of a round nose Pitot probe at low Reynolds numbers (Thompson and Holder, 1958).***

$$C_p = \frac{P_\infty - P_s}{\rho U_\infty^2/2} \tag{3.5}$$

where P_∞ is the stagnation pressure measured by the probe, P_s is the actual stagnation pressure, and $\rho U_\infty^2/2$ is the dynamic pressure measured far upstream.

3.3.4 Mach Number Effect

A Mach number variation at subsonic flow (M < 0.3) does not affect the stagnation pressure measurement significantly (Goodyer, 1975). However, shocks appearing at supersonic Mach numbers (1.0 < M < 3.0) affect the stagnation pressure measurement substantially because of the pressure losses that occur due to this shock wave at a certain distance from the measurement location. This results in readings that are higher than the stagnation pressure of the free stream.

3.3.5 Velocity Gradient Effect

The transverse velocity gradient, which is the difference in velocity between adjacent boundary layers, will result in a measurement error of the stagnation pressure measurements (Schlichting, 1979). There are two measurement errors that can result.

The first measurement error results when the pressure probes are exposed to a high stagnation pressure gradient field, which is seen when measuring inside the boundary layer or near solid boundaries. This results in the streamlines deflecting to a region of lower velocities. This deflection will cause the pressure probes to show a higher than normal stagnation pressure. A solution to minimize this measurement error is to use a flattened tip probe (called a "Boundary Layer probe" in Fig. 3.12b) to measure the stagnation pressure provided the flow direction is known.

The second error results due to the stagnation pressure being proportional to the square of the velocity. Integrating the velocity gradient at the pressure probe's nose will result in a higher stagnation pressure value than the stagnation pressure calculated from the square of the average velocity. This error can be minimized by using a "Boundary Layer probe" to reduce the distance between the center of the tube and the solid boundary surface.

3.3.6 Turbulence Effects

Turbulence or flow unsteadiness influences the pressure probes in two ways (Park, 1984). The first way is similar to that of the yaw (incidence) angle effect where the fluctuating velocity components change the direction of the flow approaching the nose of a pressure probe. The second way is that the fluctuating velocity component may contribute to the stagnation pressure. This can be seen by expressing the velocity as the sum of the mean velocity and the fluctuating components. The addition of the fluctuating components will result in change of the stagnation pressure.

Example 3.2: Aerodynamic Loss Behind a Circular Cylinder

A. Background

Flow around a blunt body such as a circular cylinder is one of the most studied in aerodynamics as it is relevant to many engineering applications (eg, heat exchangers and powerlines). To characterize the wake of flow behind the cylinder, which is associated with aerodynamic loss, the pressure field (especially total pressure field) needs to be measured. However, a precise measurement of total/static pressures cannot be implemented because the exact direction of fluid flow of the wake is essentially unknown although its nominal direction may be guessed. Therefore, it is crucial to understand how total pressure readings from various probes can mislead the loss measured due to the unknown direction of the wake.

To demonstrate the dependence of total pressure readings on the yaw (incidence) angle of the separated flow from a cylinder with respect to the probe's axis, three pressure probes shown in Fig. 3.5: (a) cylinder-type probe, (b) Pitot probe, and (c) Kiel probe, are tested.

B. Test Setup

An upright single cylinder (length to diameter ratio of 2.5) with both ends attached to endwalls is positioned inside a rectangular wind tunnel as shown in Fig. 3.8. Dimensions of a rectangular test section are $H \times W = 0.1$ m (height) $\times$ 0.4 m (width) and the cylinder diameter is $D = 0.04$ m. Prior to the experiments, the mainstream velocity was measured using a Pitot tube placed $4D$ upstream of the cylinder and fixed at 15 m/s, providing a Reynolds number of roughly 32,500.

Stagnation pressures have been measured at two selected downstream traverse planes ($x/D = 2.0$ and 10.0) using the cylinder, Pitot, and Kiel-type probes as illustrated in Fig. 3.8. The pressure probes are aligned with the nominal flow direction, the x-axis in the midspan plane. The probes are positioned such that they will measure the stagnation pressure profile in the y-axis. In addition, a separate stagnation tube is positioned $2.5D$ upstream from the cylinder to monitor the reference stagnation pressure.

Fig. 3.9 shows velocity profiles along the z-axis (channel height) at upstream traverse planes ($x/D = (-)4.0$) on the plane of symmetry where velocities are normalized by the centerline velocity U_∞ for the empty channel case. Data from only the lower half of the channel height are displayed

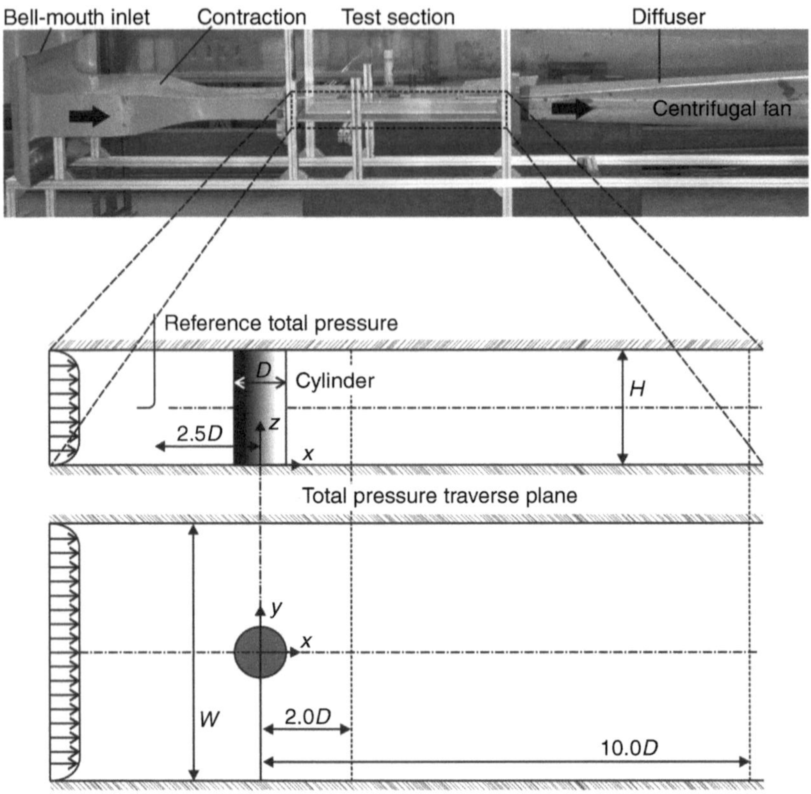

Figure 3.8 ***Schematic of experimental setup for the measurement of total pressure loss behind a circular cylinder.***

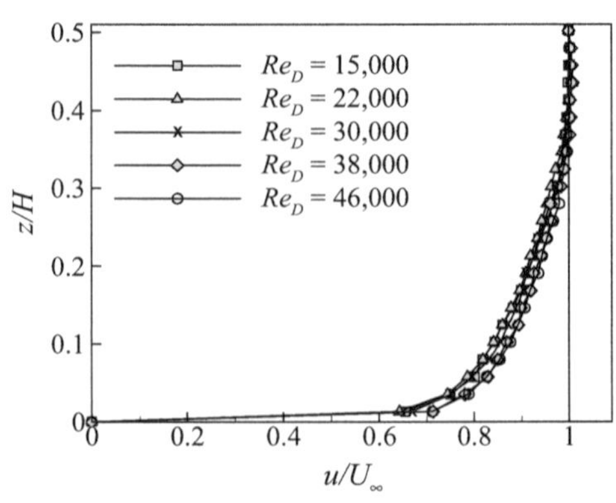

Figure 3.9 ***Upstream velocity profiles along the channel height (the z-axis) measured at x/D = 4.0 and the channel's mid-width for five selected Reynolds numbers.***

in the figure since the velocity profile was confirmed to be symmetric with respect to the midspan of the cylinder.

The stagnation pressure loss coefficient (Y_p) was defined as:

$$Y_p = \frac{p_s - p_t}{\rho U_\infty^2/2} \tag{3.6}$$

where p_t is the stagnation pressure measured by the probe, and p_s and $\rho U_\infty^2/2$ are the stagnation pressure and dynamic pressure measured far upstream, respectively.

C. Discussion of Results

The loss coefficient distribution results for the three selected stagnation probes at various distances from the plane of symmetry (ie, the x–z plane coinciding with the center of the cylinder) are plotted in Fig. 3.10. At $x/D = 2.0$, the Pitot tube has the most loss followed by the cylinder probe, and then the Kiel probe as shown in Fig. 3.10a. There is overestimation of

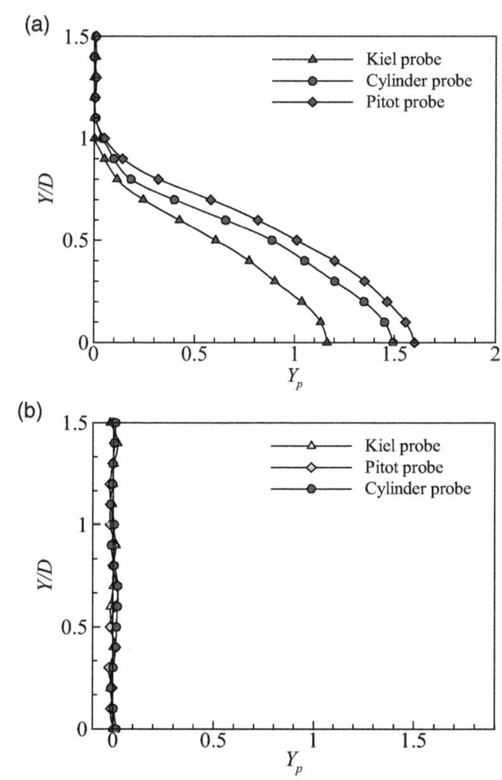

Figure 3.10 ***Loss coefficient distribution for three selected stagnation pressure probes.*** (a) At $x/D = 2.0$; (b) at $x/D = 10.0$.

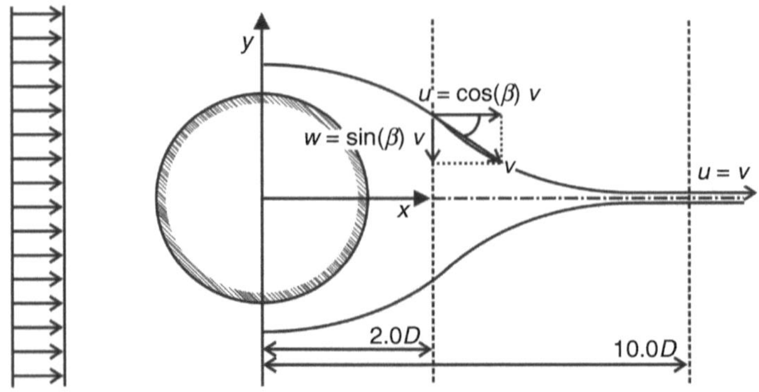

Figure 3.11 ***Velocity components of the separated flow behind a circular cylinder.***

the loss by the Pitot tube because the Pitot probe accommodates a partial component of total flow velocity. There is a flow turning by the cylinder as illustrated in Fig. 3.11. Theoretically, only the velocity component of $\cos(\beta)v$ is stagnating inside the Pitot tube where v is the actual flow velocity magnitude. This leads to a lower stagnation pressure measurement and hence a higher loss coefficient.

The Kiel probe has the least loss amongst the three probes because the Venturi-like shroud forces the local flow (v) to align itself with the shroud axis so that the flow is guided directly toward the measurement point. This ensures the u- and w-components of the flow will be measured, leading to a lower loss coefficient. However, it should be noted that there might be still an underestimation although the total pressure loss is closer to the actual value, since the actual flow angle with respect to the x-axis is unknown. It is still possible that if the Kiel probe is rotated with respect to the y-axis (or the probe's stem axis) in the z–x plane (assuming 2D flow in this plane) in an effort to find a yaw angle which gives the highest stagnation pressure, the actual stagnation (total) pressure can be measured.

At $x/D = 5.0$ in Fig. 3.10b, very low total pressure loss is observed. Three probes show the same loss profile along the y-axis and values. This indicates that the flow becomes almost parallel to the nominal flow direction and therefore the three probes measure the same total pressure.

3.4 FLOW VELOCITY MEASUREMENT

Pressure measuring devices immersed in fluid flow rather than at the boundaries are frequently called "pressure probes." When the open end of a tube faces the incoming fluid stream (its direction is assumed to be known), the flow stagnates in the tube, providing stagnation pressure there. Together with the static pressure measured in the vicinity of the position where the stagnation pressure is measured, the velocity at that position can be calculated using Bernoulli's equation as:

$$U = \sqrt{\frac{2(p_t - p_s)}{\rho}} \tag{3.7}$$

where U is the flow velocity to be calculated, p_t is the stagnation pressure, p_s is the static pressure, and ρ is the fluid density. This method was first used by Pitot in 1732 and has been extensively used ever since. An example of the Pitot tube used nowadays is shown in Fig. 3.12a.

In some cases, a stagnation tube is used alone during the velocity measurement especially for the measurement of boundary layer thickness (Fig. 3.12b). This is due to the fact that the tip of the Pitot tube (stagnation/static pressure ports) is in general rather thick compared to the momentum thickness. Therefore, the first point from the wall boundary is not close enough to

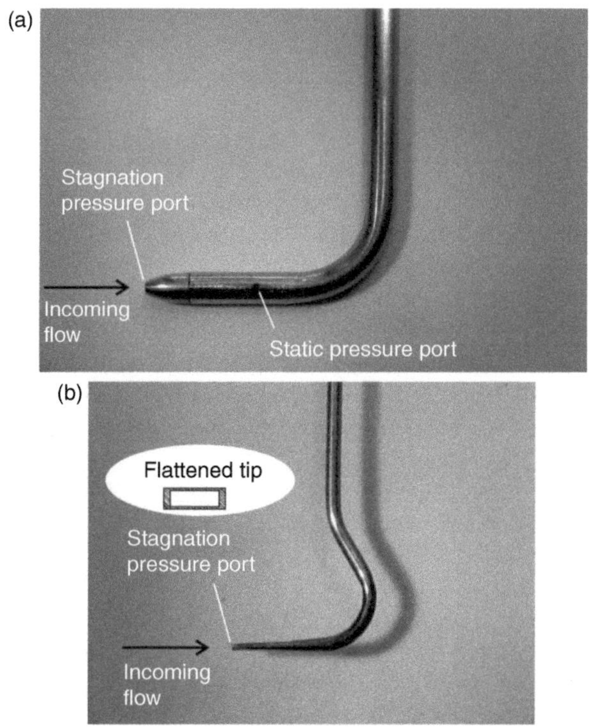

Figure 3.12 ***Examples of pressure probes for flow velocity measurement.*** (a) Pitot tube; (b) boundary layer probe.

resolve the boundary layer profile. Alternatively, the end of a circular stagnation tube is flattened to reduce the distance between the center of the tube and the solid boundary surface. In addition to the stagnation pressure (p_t), the static pressure (p_s) at the same location is required to calculate the flow velocity there. Since the stagnation tube with the flattened tip does not have a built-in static tapping unlike the Pitot tube, the static pressure has to be measured elsewhere. To this end, a static pressure tapping is inserted flush with an endwall at the same stream-wise location as the stagnation tube.

Example 3.3: Variation of the Centerline Velocity of a Free Round Jet

A. Background

Jet structures vary with the jet Reynolds number as classified by McNaughton and Sinclair (1966). Within the "semiturbulent jet" regime, $1000 < \mathrm{Re}_j < 3000$, a laminar length appears where the jet has no apparent interaction with the surrounding fluid that is seen as a "parallel-sided jet stream for a certain distance from the jet exit," termed a "laminar length." The surrounding fluid begins to entrain and to mix with the jet at the end of the laminar length. In the present round jet for $\mathrm{Re}_j = 1800$, the laminar length is estimated to be $a/D_j \sim 12.0$ as visualized in Fig. 3.13a.

At higher jet Reynolds numbers, for example, $\mathrm{Re}_j = 7500$, which falls into the "fully turbulent jet" regime, $\mathrm{Re}_j > 3000$ (McNaughton and Sinclair, 1966), the surrounding fluid is now entrained into the jet immediately after the jet exit as visualized in Fig. 3.13b. It has been established that there exist three distinctive flow regions (Abramovich, 1963): (1) the potential core region, (2) the developing region, and (3) the fully developed region. In the potential core region, the jet velocity along the jet centerline maintains its magnitude as high as that at the jet exit. Furthermore, a region where the jet is undisturbed by flow interaction with the surrounding fluid at rest is present, the so-called "potential core." Martin (1977) suggested that the potential core persists up to $L = 4D_j$ for round jets, while Gauntner et al. (1970) reported that it typically ranges from $L = 4.7D_j \sim 7.7D_j$. After the jet is fully developed via the developing region, the axial velocity profiles become self-similar (Reungoat et al., 2007) as schematically illustrated in Fig. 3.14. It is interesting to notice that both flow visualization images (Fig. 3.13a and b) show that the semiturbulent jet has a wider shear layer than that of the fully turbulent jet once entrainment takes place.

This example demonstrates how the centerline velocity of the free round jet as indicative of jet structures varies with the jet Reynolds number quantitatively. To this end, the centerline velocity of the round jet has been measured by traversing a Pitot tube along the jet axis.

(a)

(b)

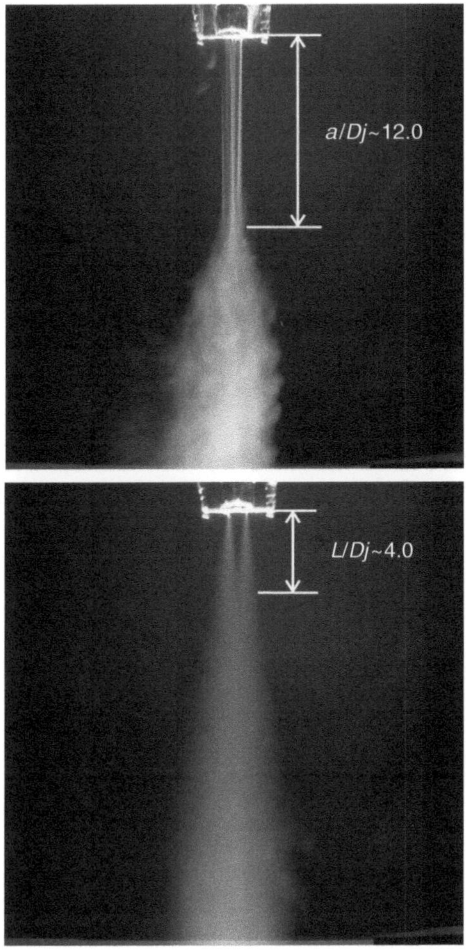

Figure 3.13 ***Jet flow pattern.*** (a) At $Re_j = 1800$ (semiturbulent jet, $L_u = a + L$); (b) at $Re_j = 7500$ (fully turbulent jet, $L_u = L$ since $a = 0$).

B. Experimental Details

B1. Test Setup

A test rig was built to investigate how the centerline velocity of a free round jet varies with the jet Reynolds number. Fig. 3.15 shows the schematic of the test rig built to measure the variation of the jet's axial velocity along the jet axis, that is, the z-axis. An air–air heat exchanger was used to cool the air supplied by a centrifugal fan, as it gets heated during the experiment (by mechanical work input from the centrifugal fan), before a settling chamber. The inner diameter of a round jet D_j is 10 mm.

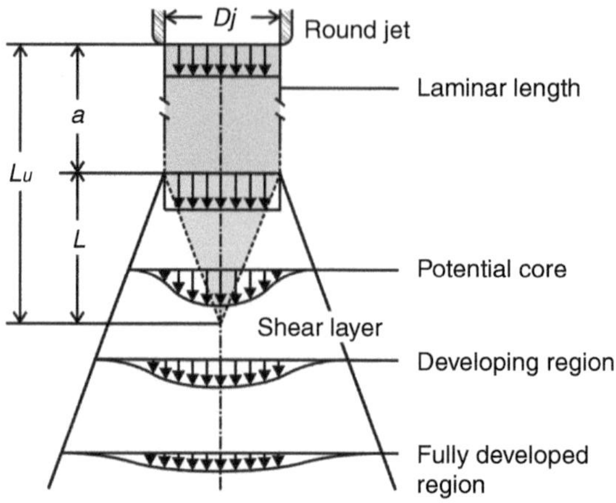

Figure 3.14 ***Schematic summary of jet flow patterns.***

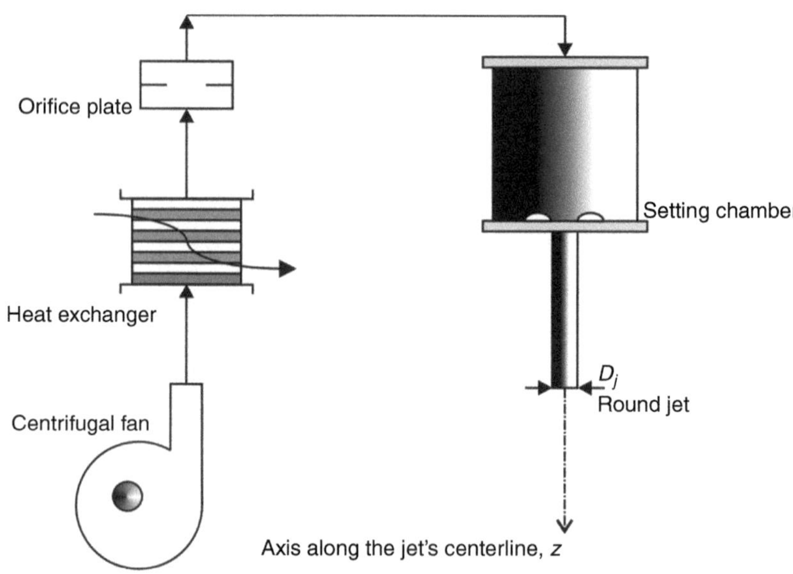

Figure 3.15 ***Schematic of the test rig used to measure the variation of a jet's axial velocity along its centerline, that is, the z-axis.***

The jet Reynolds number Re_j (based on the jet diameter D_j and the centerline velocity at a jet exit V_{jc}) was selected at Re_j = 1800, 2800, and 7500, defined as:

$$\mathrm{Re}_j = \frac{\rho w_e D_j}{\mu} \tag{3.8}$$

where w_e is the centerline jet velocity measured at the jet exit, and D_j is the inner diameter of the round jet.

To quantify jet structures, for example, laminar length and potential core of the present round jet, the centerline velocity w_0 of the jet along the jet axis in free exit was measured using a Pitot tube mounted on an automated linear traverse system. The Pitot tube diameter (1.3 mm) to round jet diameter (10.0 mm) ratio was 0.13 whilst the area blockage ratio by the Pitot tube was 1.7%. The pressure values (p_t and p_s) from the stagnation and static tappings were recorded by a differential pressure transducer (DSA, Scanivalve™).

C. Discussion of Results

The longitudinal variation of the centerline velocity of a round jet with jet Reynolds numbers covering semiturbulent jet and fully turbulent jet regimes is considered. The axial velocity component of the round jet along the jet axis (ie, the z-axis), w_0, for three selected jet Reynolds numbers, Re_j = 1800, 2800, and 7500, is considered. An "undisturbed" flow region whose length is denoted as "L_u/D_j" is defined as a downstream location where the centerline velocity w_0 reaches 98% of the jet exit velocity w_e, similar to the length of potential core (Jambunathan et al., 1992).

It is seen from Fig. 3.16 that the semiturbulent jet (Re_j = 1800 and 2800) has an undisturbed region persisting up to L_u/D_j = 15.0 (for Re_j = 1800) and L_u/D_j = 8.5 (for Re_j = 2800). It is apparent that the length of the undisturbed region decreases as the jet Reynolds number increases. The undisturbed flow region of the semiturbulent jet consists of the laminar length (a/D_j) and the length of potential core (L/D_j). The latter is invisible in the flow image in Fig. 3.13a. The length of potential core (L_p) may be estimated from the flow image (Fig. 3.13a for Re_j = 1800) to be approximately L/D_j = 3.0 [ie, the length of the undisturbed region (L_u/D_j = 15.0) – the laminar length (a/D_j = 12.0)].

Once the jet becomes fully turbulent, the laminar length disappears (Fig. 3.13b). The length of the undisturbed region is equal to that of the

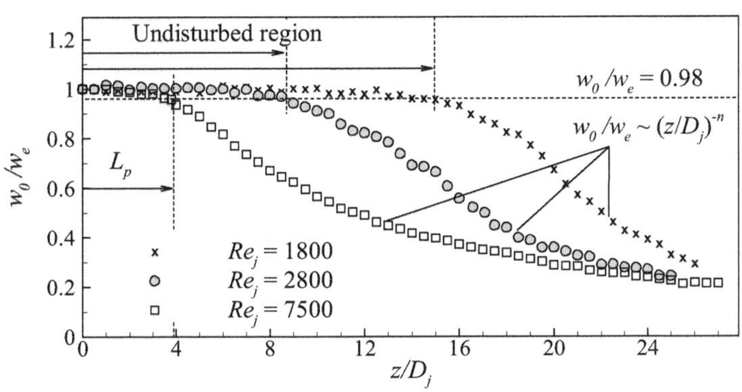

Figure 3.16 ***Variation of the centerline velocity (w0/we) of a free round jet for three selected jet Reynolds numbers.***

potential core ($L_u = L_p$), which remains unchanged regardless of the jet Reynolds number. Outside the potential core, the axial centerline velocity w_0 decays inversely proportional to the axial distance (Abramovich, 1963), as

$$\frac{w_0}{w_e} = \left(\frac{z}{D_j}\right)^{-n} \tag{3.9}$$

At, for example, $\mathrm{Re}_j = 7500$, the measured centerline velocity w_0 followed the trend expressed by Eq. (3.9) with a value $n = 1$ as proposed by Martin (1977), that is:

$$\frac{w_0}{w_e} = \left(\frac{z}{D_j}\right)^{-1} \tag{3.10}$$

However, for the semiturbulent jet at $\mathrm{Re}_j = 1800$ and 2800, the axial centerline velocity w_0 after the potential core decays inversely proportional to the axial distance but with values of $n = 2.93$ (for $\mathrm{Re}_j = 1800$) and $n = 1.8$ (for $\mathrm{Re}_j = 2800$). It is seen that the semiturbulent jet has a steeper decrease in axial centerline velocity and its extent within the semiturbulent jet regime is reduced as the jet Reynolds number is increased. With a further increase in the jet Reynolds number toward the fully turbulent jet, the longitudinal decay of w_0 is attenuated, approaching that of the fully turbulent jet, that is, $n = 1.0$. The different rates of decay of the centerline velocity may indicate a wider fanning of the shear layer of the round jet once transition takes place.

3.5 FLOW-DIRECTION MEASUREMENT

3.5.1 Background

For the precise description of a flow field, both the flow direction and velocity magnitude are needed. The detailed flow direction data allow better alignment of a probe to the flow as required for accurate total and static pressure measurements, leading to accurate velocity magnitude and aerodynamic loss data.

Variations of the velocity component in the plane perpendicular to the probe stem are defined as the yaw angle. Variations of the velocity direction in the plane perpendicular to it are defined as the pitch angle. This section deals with pneumatic probes that allow flow direction measurements for the yaw angle only (2D flow field) and the yaw and pitch angles (3D flow field), separately.

3.5.2 Three-hole Probe

A three-hole probe is commonly used to determine the direction of flow, enabling the yaw angle to be calculated from the measured pressures. The

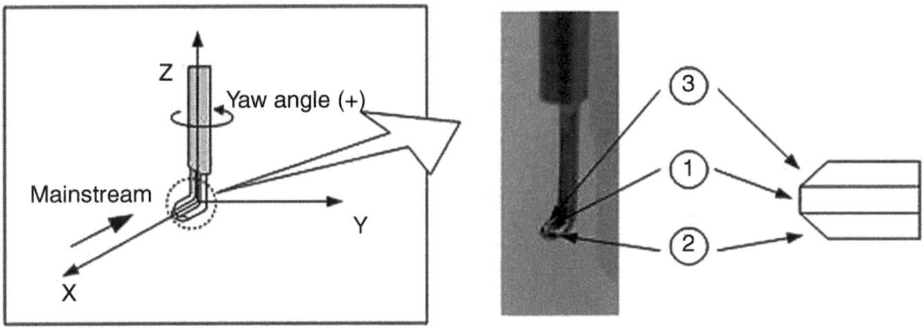

Figure 3.17 ***Configuration for yaw angle and a photograph of a three-hole probe (from United sensor™).***

coordinates defining the axes of the yaw angle for a particular type (Cobra type) of three-hole probe are given in Fig. 3.17. Relative to the center pressure port (1), the two outside ports (2, 3) have their forward tips cut off at 45°. All three tubes lie on a single plane (x–y plane). This section describes the details of calibration setup, procedure, and typical results of flow measurements using the three-hole probe.

3.5.2.1 Parameters

During calibration, pressures p_1, p_2, and p_3 from three pressure ports of the three-hole probe are acquired where the subscripts indicate the corresponding pressure port numbers as illustrated in Fig. 3.17. In common practice, the pressure expressed in nondimensional form, that is, the pressure coefficient (Cp), is used. For yaw angle, which is expected to be dependent upon the pressures measured at pressure ports 2 and 3, the pressure coefficient of the yaw angle may be defined as (Hodson, 1996):

$$Cp_{yaw} = \frac{p_2 - p_3}{p_1 - p_{avg}} \tag{3.11}$$

where p_{avg} is the mean of the measured pressures at two pressure ports other than the centered port, defined as $p_{avg} = (p_2 + p_3)/2$. Similarly, the pressure coefficient at each hole may be defined as:

$$Cp_i = \frac{p_i - p_s}{p_t - p_s} \tag{3.12}$$

where p_t is stagnation pressure, p_s is static pressure, and p_i is the pressure at each hole ($i = 1, 2, 3$). Eq. (3.11) is used to obtain Cp_{yaw} from flow measurements.

3.5.2.2 Calibration

For calibration, the three-hole probe is mounted on a calibration unit consisting of a three-axis rotary table, a free-jet generator, and a data acquisition system (Fig. 3.18a and b). The free-jet generator was connected to a computer-controlled module. The jet velocity was controlled and monitored by a personal computer. The typical pattern of the radial velocity profile at a given distance L from the nozzle exit is shown in Fig. 3.19a and b.

A three-hole probe (head diameter 3 mm) was used. The principal advantage of this type of three-hole probe over conventional pressure probes is less flow disturbance due to the presence of the probe (ie, the potential effect). The calibration of the probe is performed with a free-jet. The probe is placed at the exit of the free-jet generator where the flow is injected through a smooth isentropic nozzle. Special care is required to set a reference position for zero yaw angle. To this end, an analogue pressure transducer was used to

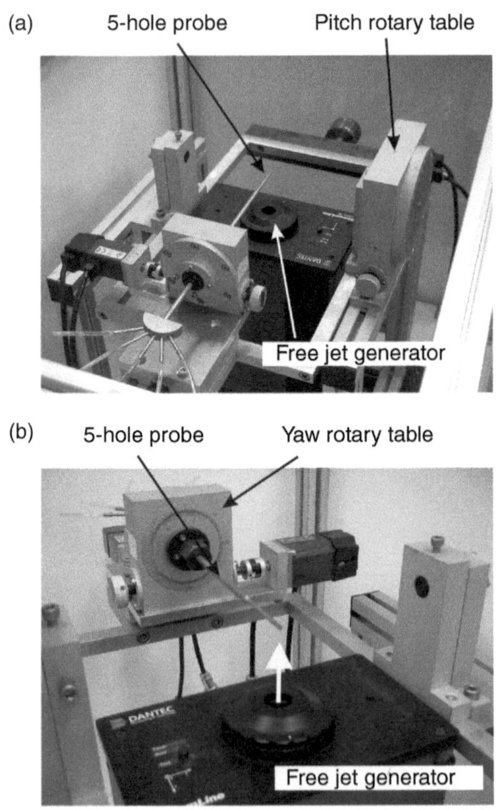

Figure 3.18 ***Calibration setup including a free-jet generator and rotary tables.*** (a) For pitch angle; (b) For yaw angle.

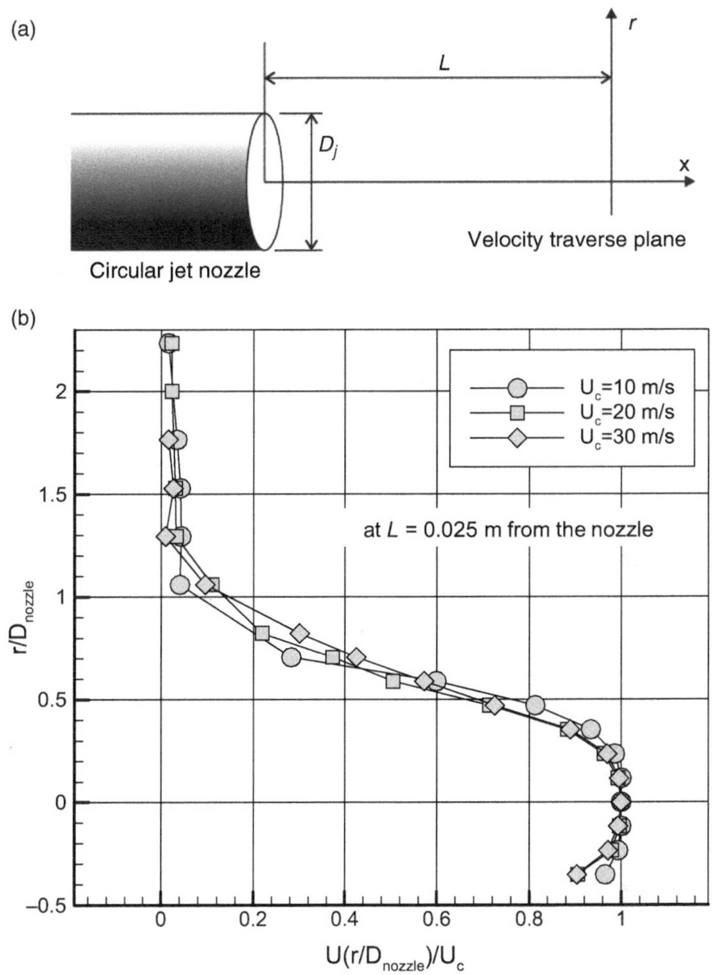

Figure 3.19 ***Characteristics of the free-jet generator used in the present study.*** (a) Coordinates of jet and traverse plane; (b) radial velocity profile at a fixed distance, $L = 0.025$ m.

manually set the reference position. At the ideal reference position in the presence of the free-jet, the pressure at port 1 indicates the maximal value while those at port 2 and port 3 for yaw angle are identical, respectively.

The calibration is carried out by varying the yaw angle after the reference position is determined. In the present study, the yaw angle is varied from −45 to +45° with a 3° increment. In addition, three different free-jet velocities (5, 10, and 20 m/s) at the jet core region were tested to examine the dependence of the yaw angle on free-jet velocity.

3.5.2.3 *Calibration Results and Discussion*

3.5.2.3.1 Dependence of Pressures at Each Port on Yaw Angle

The incidence angle of the free-jet to each port is different for different yaw angles. As the yaw angle is varied, the raw pressure coefficients at different ports have different responses as seen in Fig. 3.20a, which plots the pressure coefficient from the centered port, p_1, as a function of the yaw angle. In general, the pressure appears to be symmetric with respect to the reference line (ie, zero yaw lines), as expected. For both positive and negative yaw regimes, the pressure monotonically decreases as the yaw angle is increased. For ports 2 and 3, the pressure response at each port shows an inverse trend as the yaw angle is varied.

Fig. 3.20b shows the dependence of Cp_{yaw} on yaw angle. The curves corresponding to different free-jet velocities are almost symmetrically centered on the origin. As the yaw angle is increased, the magnitude of Cp_{yaw} monotonically increases.

3.5.2.3.2 Determination of Yaw Angle

During the experiments, the yaw angle needs to be determined based on the measured pressure values at three pressure ports. Fig. 3.20b plots the pressure coefficients at the three different free-jet velocities as functions of the yaw angle. There exists a unique value of the yaw angle for a given value of Cp_{yaw}, irrespective of the free-jet velocity. In application of the three-hole probe for flow direction measurements, Cp_{yaw} values are firstly obtained and then the yaw angle can be inversely determined from the contour maps provided in Fig. 3.20b.

3.5.3 Five-hole Probe

A five-hole probe is commonly used for determining the direction of flow, yaw angle, and pitch angle from measured pressures. The coordinates defining the axes of the yaw and pitch angles for a particular type of five-hole probe are given in Fig. 3.21a. This section describes the details of calibration setup, procedure, and obtained results with five-hole probes.

3.5.3.1 *Parameters*

During the calibration, pressures p_1, p_2, p_3, p_4, and p_5 from five different pressure ports of the five-hole probe are acquired, as illustrated in Fig. 3.21b.

For the yaw angle, which is expected to be mainly dependent upon the pressures measured at pressure ports 2 and 3, the pressure coefficient may be defined (Hodson, 1996) as:

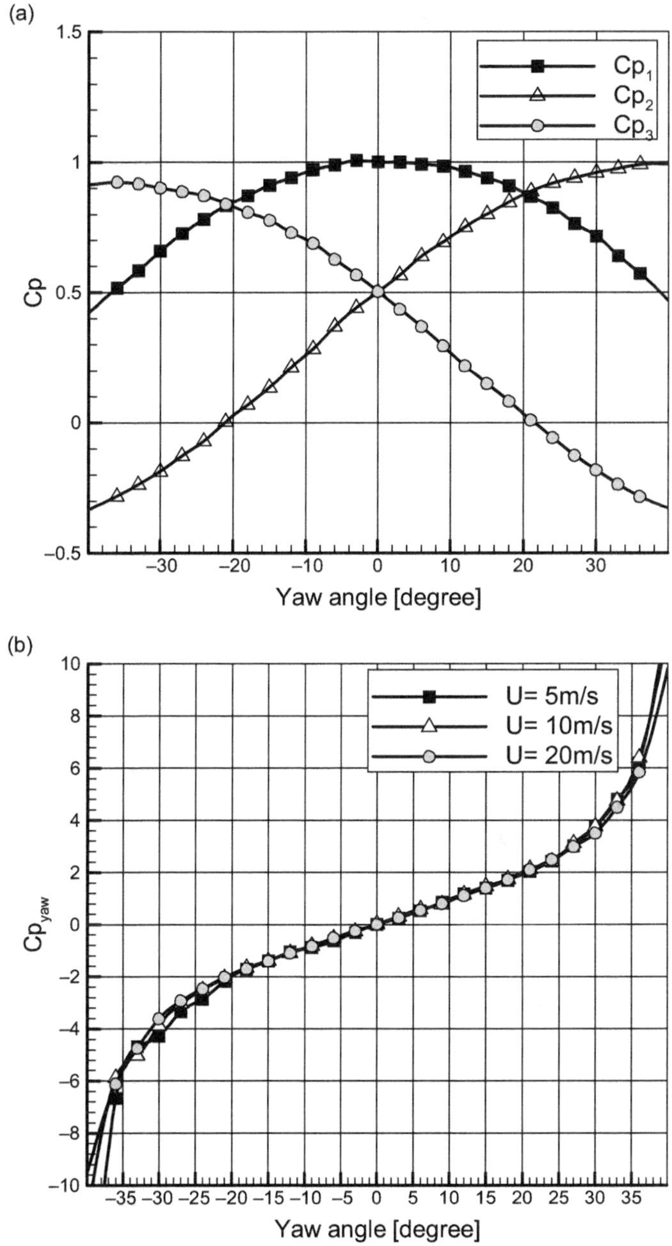

Figure 3.20 ***Dependence of Cpyaw on: (a) yaw angle; and (b) yaw angle and flow velocity.***

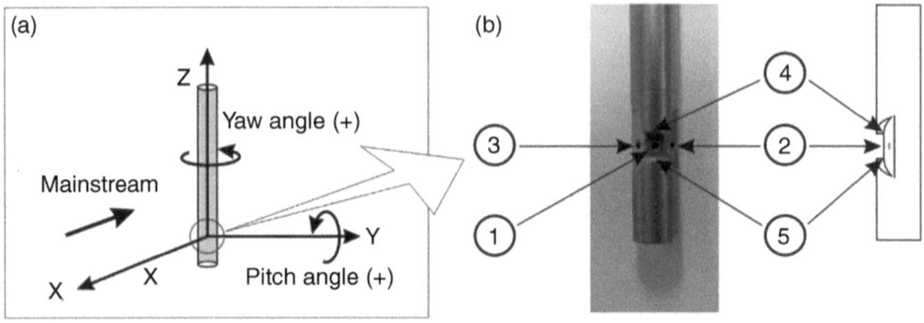

Figure 3.21 (a) Configuration for yaw and pitch angles and (b) photograph of an upright-type five-hole probe.

$$Cp_{yaw} = \frac{p_2 - p_3}{p_1 - p_{avg}} \tag{3.13}$$

where p_{avg} is the mean of the measured pressures at four pressure ports (other than the centered port), defined as $p_{avg} = (p_2 + p_3 + p_4 + p_5)/4$. Similarly, for the pitch angle, the pressure coefficient may be defined as:

$$Cp_{pitch} = \frac{p_4 - p_5}{p_1 - p_{avg}} \tag{3.14}$$

During the flow measurements, both Cp_{yaw} and Cp_{pitch} are to be obtained. The corresponding yaw and pitch angles of the flow are determined after matching with the data map of Cp_{yaw} and Cp_{pitch}. Details of how this inverse matching is performed to extract the yaw and pitch angles are presented next.

3.5.3.2 Calibration

For calibration, the five-hole probe is mounted on a calibration unit consisting of three-axis rotary tables, a free-jet generator, and a data acquisition system. The free-jet generator was connected to a computer-controlled module. The jet velocity was controlled and monitored by a personal computer.

A five-hole probe (head diameter 3 mm) was used. It should be noted that the configuration of this probe differs from that of conventionally used probes as seen in Fig. 3.21b. All of the pressure ports are located on the side surface of the cylindrical tube near the tip (ie, an upright type). Due to the unique configuration of the probe, it is expected that the measurement range of the pitch angle is limited in comparison with that of the yaw angle. The principal advantage of this type of five-hole probe over other probes is its applicability for more compact flow measurements. Typical Cobra-type probes need space at least for the bent neck of the probe. On the other hand, when the flow behind an object open to a large space such as flow behind linear cascades is of interest, the Cobra-type probe is a better choice since less flow disturbance due to the presence of the probe is expected.

The calibration of the probe is performed with a free-jet. The probe is placed at the exit of the free-jet generator where the flow is injected through a smooth isentropic nozzle. Special care is taken to set the reference position for zero yaw and pitch angles, with an analogue pressure transducer employed to manually set the reference position. For an ideal reference position, the pressure at port 1 is at the maximum while those at ports 2 and 3 for yaw angle and those at ports 4 and 5 for pitch angle are identical, respectively.

Similar to that for the three-hole probe, the calibration is carried out with varying yaw and pitch angles after the reference position is found. In this study, the yaw and pitch angles range from −45 to +45° with a 3° increment. In addition, two different Reynolds numbers ($Re_d = 1.0 \times 10^3$ and 2.0×10^3) at the jet core region were tested to examine the dependence of these angles on the velocity.

3.5.3.3 Calibration Results and Discussion

3.5.3.3.1 Dependence of Pressure at Each Port on Yaw Angle and Pitch Angle

The incidence angle of the free-jet flow to each port is different for the set values of the yaw and pitch angles. As the yaw and pitch angles are varied, the raw pressures at different ports have different responses as seen in Fig. 3.22a–e at $U = 20$ m/s.

The pressure at the centered port p_1 is plotted as a function of the yaw and pitch angles in Fig. 3.22a. In general, the pressure is symmetric with respect to the reference lines, that is, zero yaw and zero pitch lines. For the positive yaw and pitch regions, the pressure monotonically decreases as the angles are increased. For ports 2 and 3, the pressure response at each port shows an inverse trend relative to the yaw angle as presented in Fig. 3.22b and c. Similarly, for ports 4 and 5, the pressure has an inverse trend with respect to the pitch angle as shown in Fig. 3.22d and e.

Fig. 3.23 displays the dependence of Cp_{yaw} and Cp_{pitch} on the yaw (Fig. 3.23a) and pitch angles (Fig. 3.23b), respectively, with $U = 20$ m/s. The pressure coefficient of the yaw angle appears to follow a third-order power curve as shown in Fig. 3.23a. The curves are almost symmetrically centered on the origin regardless of the pitch angle. For positive yaw angles, Cp_{yaw} monotonically increases in magnitude with increasing yaw angles; the same trend holds for negative yaw angles. In addition to the yaw angle, Cp_{yaw} is also dependent on the pitch angle especially at relatively large yaw angles.

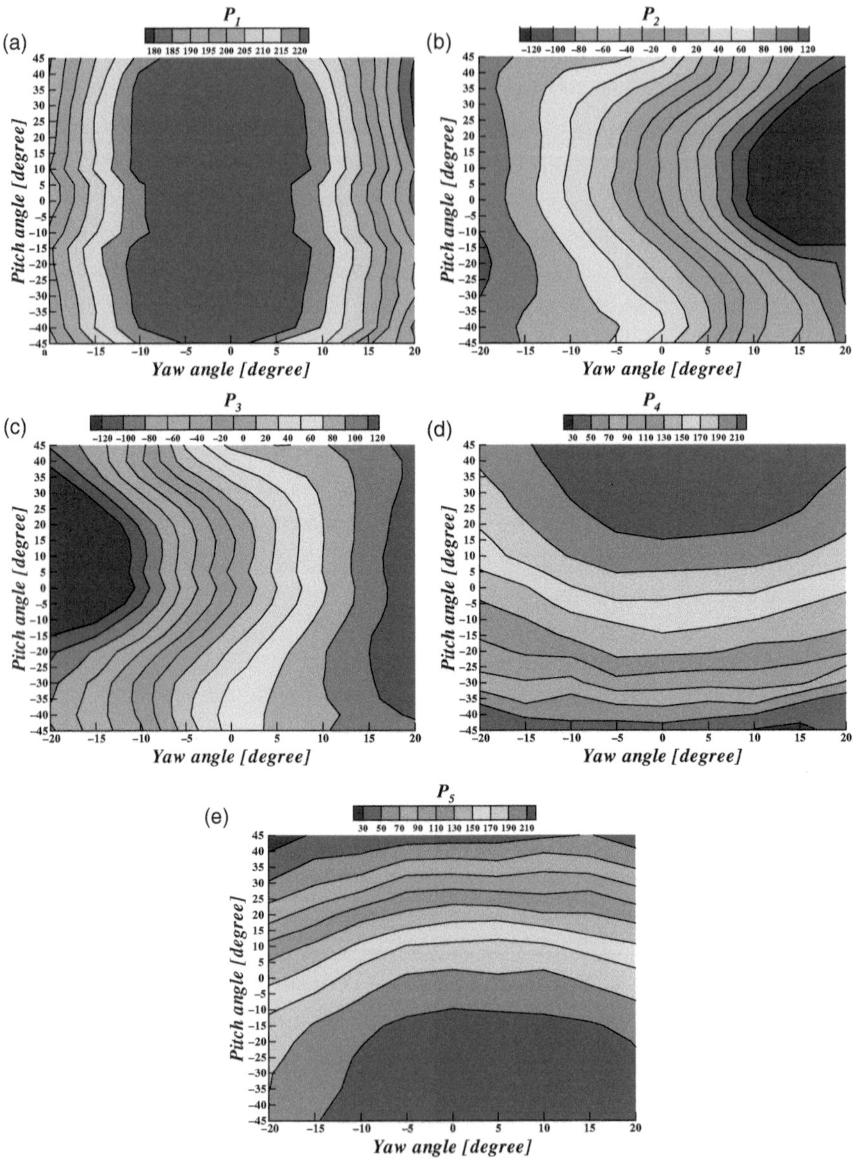

Figure 3.22 ***Raw pressure at each port measured in Pascal at U = 20 m/s.*** (a) Port 1; (b) port 2; (c) port 3; (d) port 4; (e) port 5.

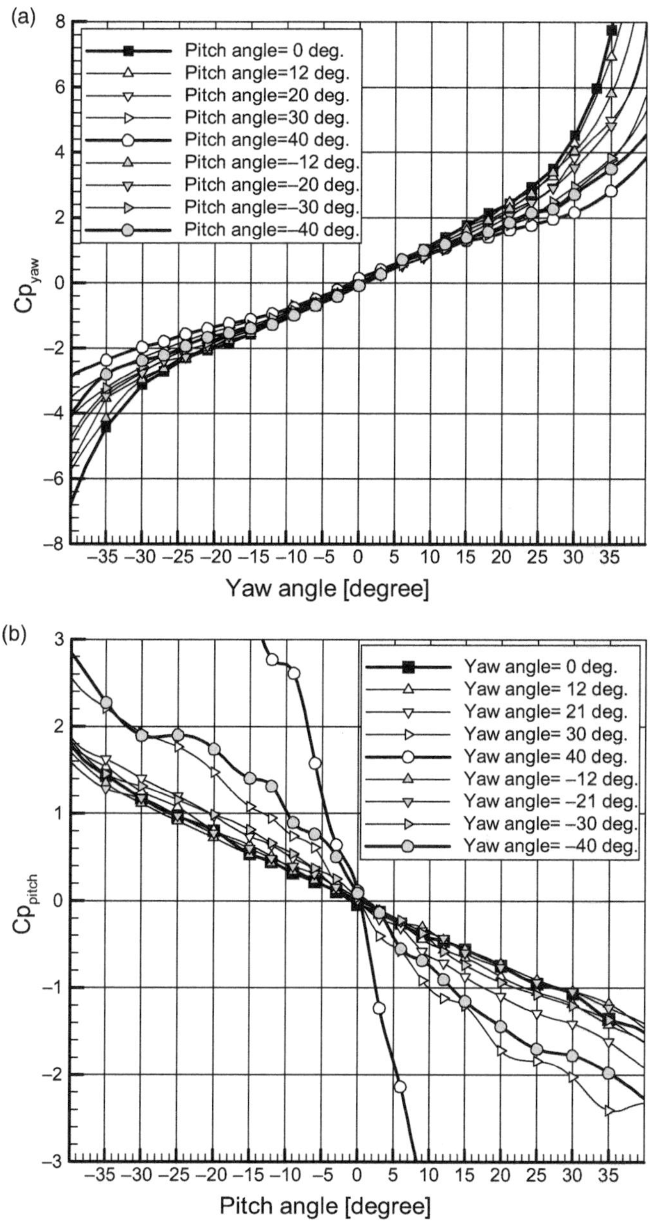

Figure 3.23 ***Dependence of Cpyaw on: (a) yaw angle; and (b) pitch angle for fixed free-jet core velocity (U = 20 m/s).***

The variation of Cp_{pitch} with pitch angle is presented in Fig. 3.23b for selected yaw angles. Unlike the yaw angle case, the pressure is strongly dependent upon the sign of the pitch angle when the angle is relatively large, for example, ±40°. This implies that a manufacturing imperfection may exist in the probe, leading to different flow patterns formed at large conjugated pitch angles. However, when the pitch angles are less than about ±30°, the flow is symmetric with respect to the origin (zero pitch angle).

3.5.3.3.2 Effect of Reynolds Number on Yaw and Pitch Angles

This section examines the influence of the Reynolds number on the yaw and pitch angles. Two selected Reynolds numbers ($Re_d = 1.0 \times 10^3$ and 2.0×10^3, based on the probe diameter, *d*) which correspond to the jet velocities 5 and 10 m/s were tested, and the results are plotted in Fig. 3.24a and b where the dotted and solid curves indicate the cases of $Re_d = 1.0 \times 10^3$ and 2.0×10^3, respectively. In general, the pressure data obtained from the two different experiments show similar responses of Cp_{yaw} and Cp_{pitch} to the yaw and pitch angles, regardless of the imposed jet velocity. However, as both angles exceed ±25°, the influence of the velocity (or Reynolds number) on both pressure coefficients becomes obvious. In other words, if the flow angles are within ±25°, the Reynolds number effects on pressure coefficients may be assumed to be negligible (Hodson, 1996).

3.5.3.3.3 Determination of Yaw Angle and Pitch Angle

During the experiments, the yaw and pitch angles need to be determined based on the measured pressure values at five pressure ports. As noticed previously, yaw and pitch angles simultaneously influence Cp_{yaw} and Cp_{pitch}. Therefore, the yaw and pitch angles must be determined based on correlation as functions of both pressure coefficients (not based on a single pressure coefficient correlation).

Fig. 3.25a and b show contour maps for the yaw angle and pitch angle, respectively. There exists a unique value of yaw and pitch angles for given values of Cp_{yaw} and Cp_{pitch}. In the applications of the five-hole probe for flow direction measurement, Cp_{yaw} and Cp_{pitch} are obtained and then the yaw and pitch angles can be inversely determined from the contour maps provided in Fig. 3.25.

Example 3.4: Flow Turning by an Inclined Circular Cylinder

A. Background

In many engineering applications, a single cylinder with both ends attached to endwalls is positioned at a certain angle of attack (inclination) to the

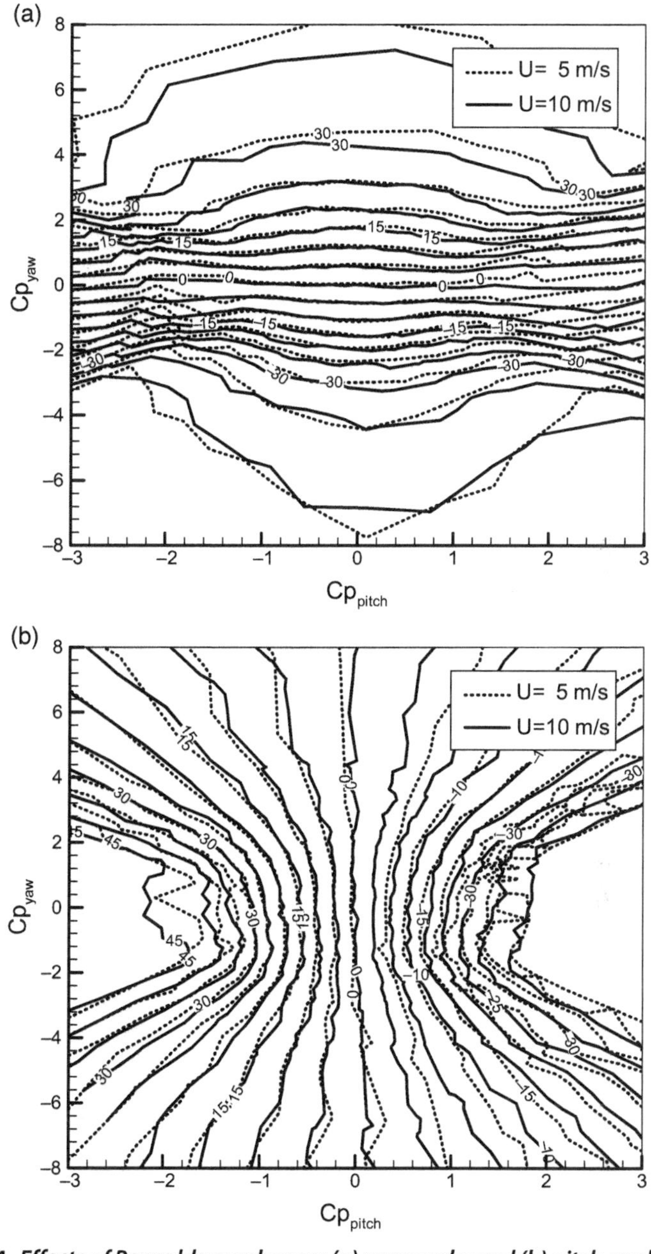

Figure 3.24 ***Effects of Reynolds number on: (a) yaw angle; and (b) pitch angle.***

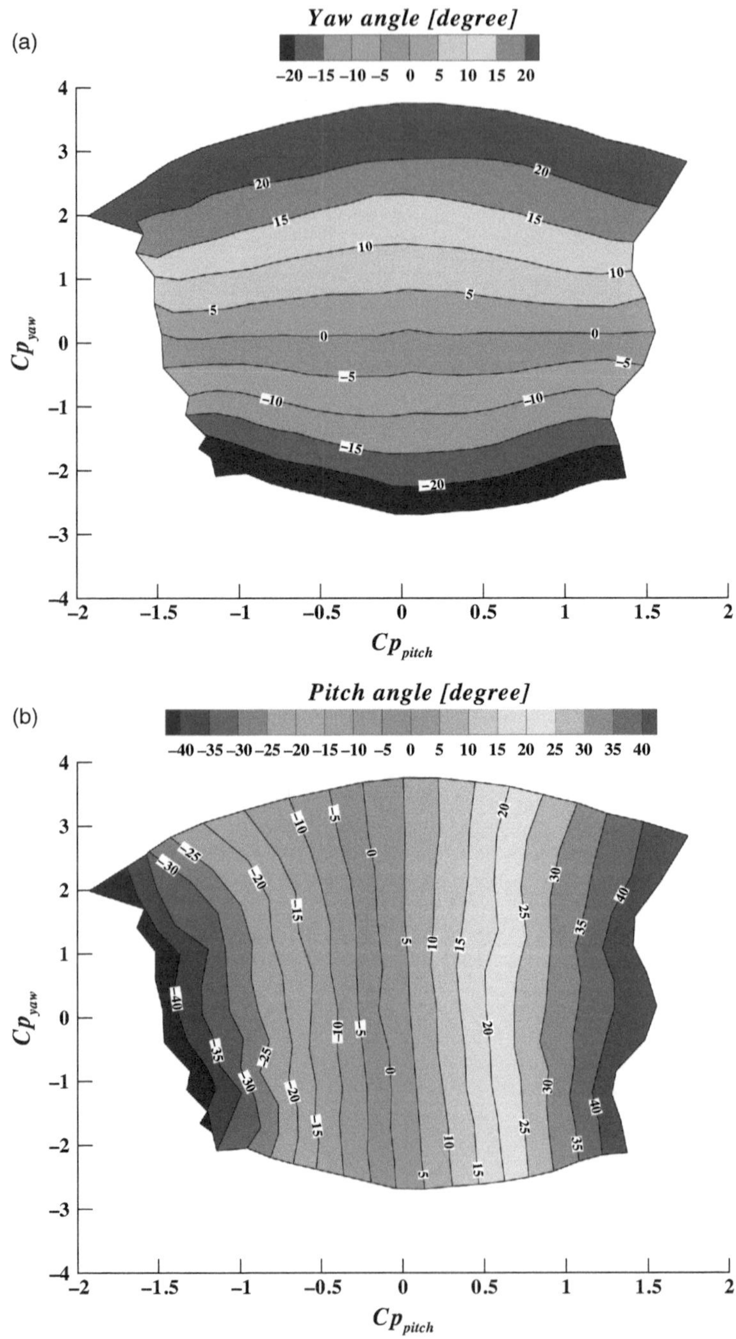

Figure 3.25 ***Determination of: (a) yaw angle and (b) pitch angle from Cppitch and Cpyaw data.***

Figure 3.26 ***Steam turbine rotor blades connected by circular tie-rods.***

mainstream flow (or the flow approaches the cylinder at an inclined angle). Fig. 3.26 shows high aspect ratio rotor blades which are connected by circular tie-rods to reduce vibration during operation. In such configurations, vortical flow patterns around the cylindrical rods become more complex due to new flow features induced by inclination. According to Choi et al. (2007), the inclination-induced flow features around a cylinder include: (1) skewed upstream main flow toward the endwall where low pressure exists as the flow approaches the cylinder; (2) formation of a "jet-like flow" immediately downstream of the cylinder, followed by its impingement onto a region of the endwall downstream of the cylinder junction; and (3) skewed separation line along the cylinder span from the cylinder axis. Amongst these features, the skewed mainstream and jet-like flow may influence horseshoe vortices formed at the junction.

B. Test Setup

An open-type wind tunnel was used for flow angle measurement downstream of the cylinders with/without inclination. Air drawn from a centrifugal fan passes through a settling chamber, screens, and a contraction section. A schematic of the test section is illustrated in Fig. 3.27a and b.

The time-averaged flow angles were measured at selected downstream traverse planes (x/D = 2 and 4) using a five-hole probe (head diameter 3 mm). The stagnation pressure port of the five-hole probe was aligned with the nominal flow direction, that is, x-axis, for the selected y–z traverse planes. The probe was mounted on an automated 2D traverse system so that velocity vectors could be obtained in each traverse plane. Pressure tubes of the probes were connected to a 48-port J-type Scanivalve™. In addition, a Pitot tube was positioned 2.5D upstream from the cylinder to monitor the mainstream velocity.

C. Discussion of Results

C1. Flow Field Behind an Upright Single Cylinder

At midspan in the x–y plane, the wake shedding patterns including the width of alternating wakes are sensitive to the inclination of the cylinder.

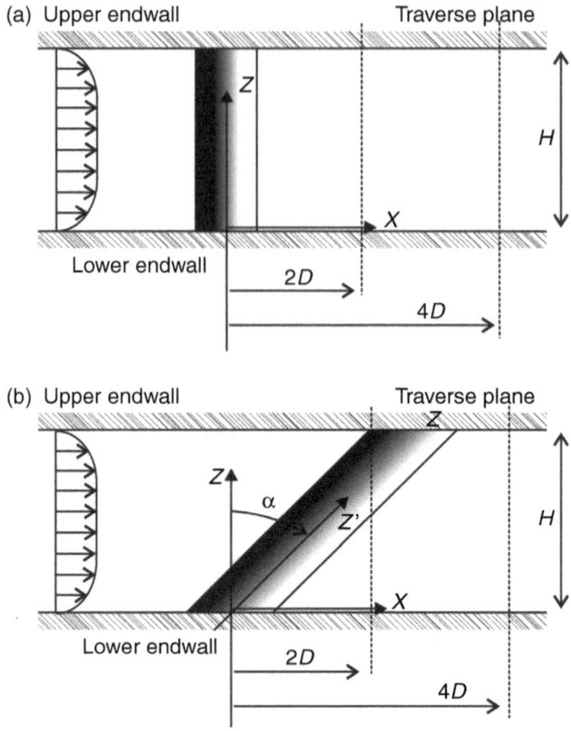

Figure 3.27 ***Schematic of (a) upright cylinder and (b) inclined cylinder along the direction of flow, with the ends of the cylinder attached to flat plates (endwalls).***

The mainstream flow is also turned pitchwise, or skewed (in the z–x plane), by the inclined cylinder. This skewed mainstream flow interacts with the lower/upper endwalls. To examine how the presence of the inclined cylinder affects the flow downstream of the cylinder, the flow angles and velocities were measured at two selected downstream traverse planes, $2D$ and $4D$ from the cylinder axis on the lower endwall.

Fig. 3.28a presents the spanwise variation of the area-averaged pitch angles for the upright cylinder case, covering a transverse range near the cylinder, that is, $(-)1.5 \leq y/D \leq (+)1.5$, for each selected spanwise location (along the z-axis) where only the lower half of the cylinder span is considered. Near the lower endwall, the mainstream shows negative pitch angles. In this study, a positive sign of the pitch angle denotes flow directing downward to the lower endwall. Due to the presence of boundary layers on endwall surfaces (not quantified here), the flow is displaced from the endwall. Thus, streamlines near the lower endwall are directed toward the midspan, showing the negative pitch angles. Moving away from the lower endwall, the flow pitch angle decreases.

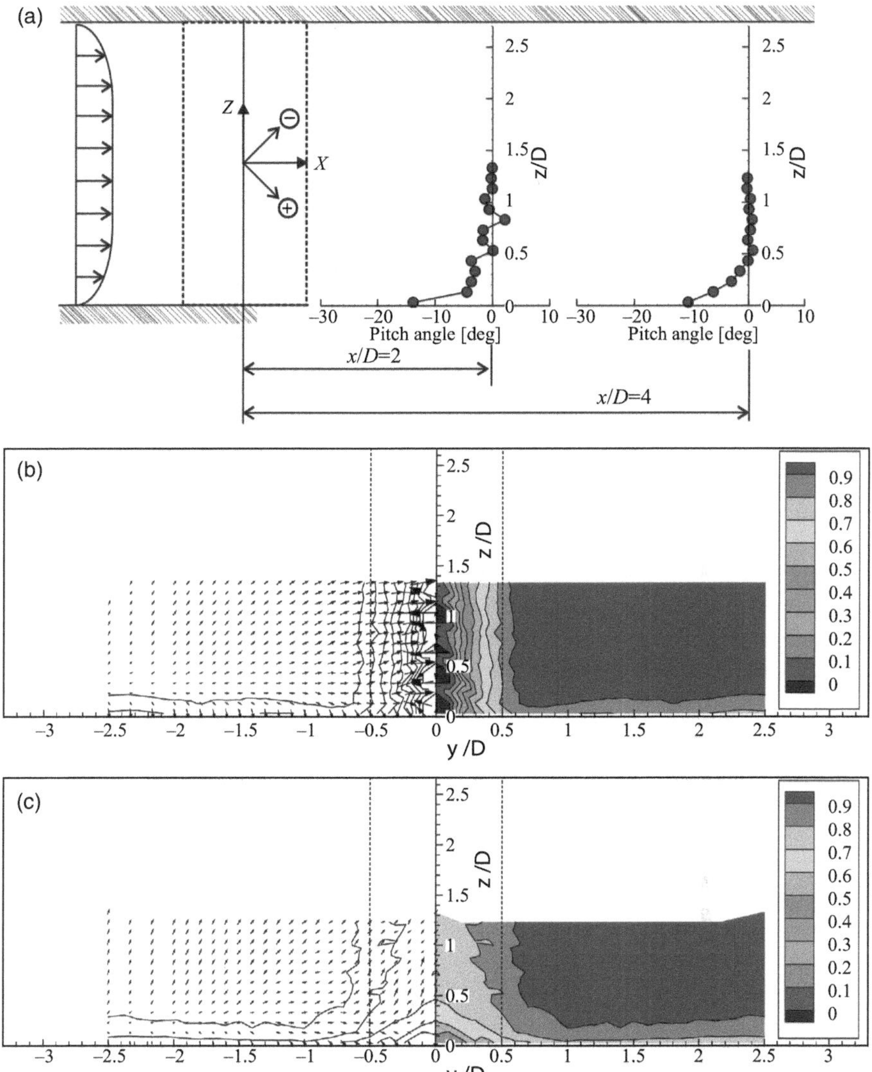

Figure 3.28 ***Flow properties behind an upright cylinder at*** $Re_D = 1.5 \times 10^4$**.** (a) Spanwise variation of the area-averaged pitch angles; (b) *y–z* plane velocity vectors and axial velocity contours at $x = 2D$; (c) plane velocity vectors and axial velocity contours at $x = 4D$.

The mainstream flow in the range of $z/D \sim 0.5$ to 1.35 (midspan) has 0° pitch angle.

Fig. 3.28b and c show separately the *y-z* velocity vectors (left side) and the contour of axial velocity normalized by the inlet mainstream velocity (right side) at the $x/D = 2.0$ plane. A large gradient of the axial velocity is concentrated near the cylinder edge ($y/D = \pm 0.5$) along the cylinder span where the flow separates. The y-z velocity vectors also indicate that the flow is directed toward the cylinder axis, that is, $y/D = 0$ due to the vorticity filament formed parallel to the cylinder axis. In addition, there is a negative axial velocity region aligned with the cylinder axis displayed as zero velocity. With the present five-hole probe, the negative velocity region could not be measured. This region implies a recirculating flow. Another velocity gradient observed near the endwall is due to the wall boundary layer. However, its gradient is not as large as that aligned with the cylinder span.

C2. Flow Field Behind an Inclined Single Cylinder

The mainstream flow is skewed by an inclined cylinder. To quantify flow turning, the pitch angles have been measured for a cylinder with 45° inclination, and the corresponding results are plotted in Fig. 3.29a. The flow at $0 \leq z/D \leq 1.5$ is under the direct influence of the separated flow from the cylinder while that at $1.5 \leq z/D \leq 2.67$ is passing by the side of the cylinder, where $z/D = 0$ and 2.67 denote the locations of the lower and upper endwalls, respectively. For the same reason as in the upright cylinder case, the flow near the lower endwall shows negative pitch angles. Away from the lower endwall, the flow has a positive pitch angle, heading toward the lower endwall at $0.1 \leq z/D \leq 0.8$. When the mainstream flow separates from the cylinder surface, it is turned, resulting in a positive pitch angle. However, at $z/D = 0.8$, there exists a sudden change in the sign of the pitch angle, from positive to negative, due to the presence of the jet-like flow immediately downstream of the cylinder, at $0.8 \leq z/D \leq 1.1$. The velocity vectors shown in Fig. 3.29b clearly indicate the formation of two counter-rotating vortices. The mainstream, then, flows parallel to the endwalls before its angle gradually becomes positive, which is associated with the separated flow from the leading edge of the upper endwall vertex.

At $x/D = 4$ (Fig. 3.29c), the flow in the lower half near the lower endwall flows in general downward (positive pitch angle) while that in the upper half flows toward the upper endwall. A very large gradient of the pitch angle is clearly seen at $1.5 \leq z/D \leq 2.2$. This region is associated with the approach of the jet-like flow toward the upper endwall. Clearly, two counter-rotating vortices are visible. The pitch angles increase, resulting from the interaction with a horseshoe vortex formed around the upper endwall. The low axial momentum vortical core of the horseshoe vortex acts to reduce the pitch angle of the flow near the upper endwall. The jet-like flow accumulates behind the upper part of the cylinder and convects downstream. The axial velocity behind the lower cylinder part has fully recovered to the mainstream value.

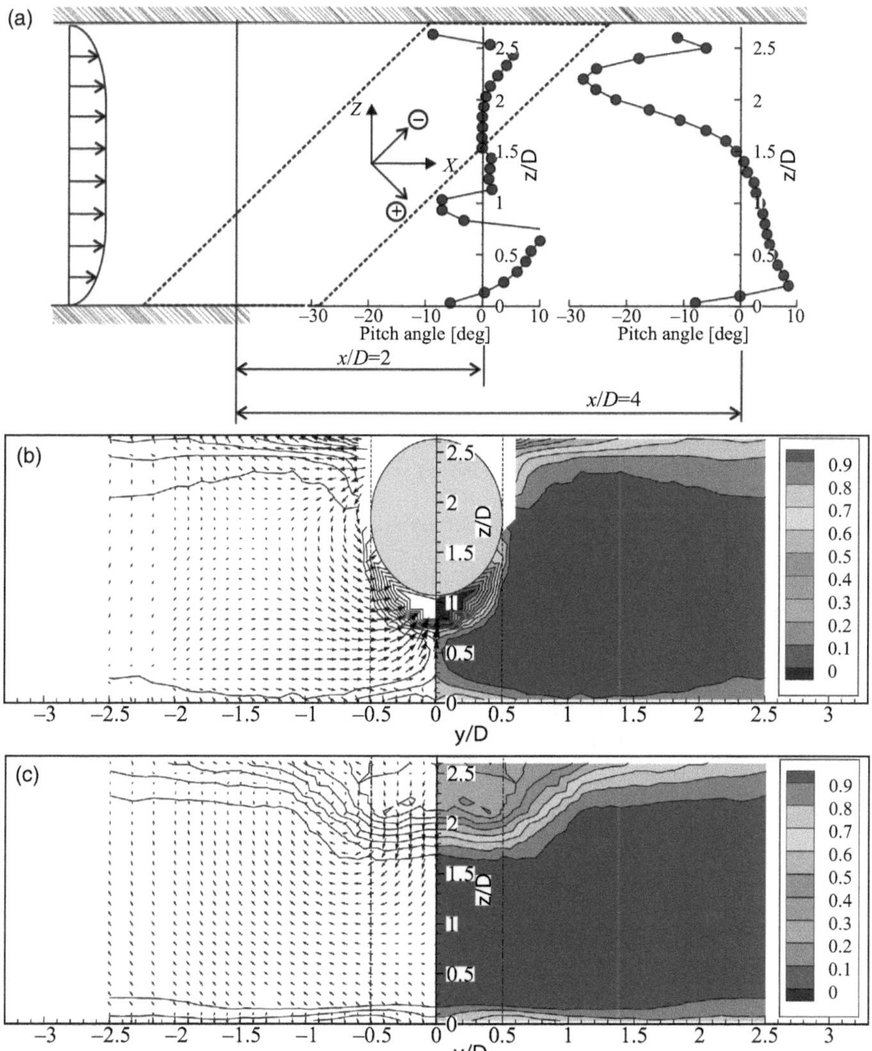

Figure 3.29 ***Flow properties behind an inclined cylinder at $Re_D = 1.5 \times 10^4$.*** (a) Spanwise variation of the area-averaged pitch angles; (b) *y*–*z* plane velocity vectors and axial velocity contours at $x = 2D$; (c) *y*–*z* plane velocity vectors and axial velocity contours at $x = 4D$.

REFERENCES

Abramovich, G.N., 1963. The Theory of Turbulent Jets. MIT Press, Cambridge, MA.

Choi, I.K., Kim, T., Song, S.J., Lu, T.J., 2007. Endwall heat transfer and fluid-flow around an inclined short cylinder. Int. J. Heat Mass Tran. 50, 919–930.

Gauntner, J.W., Livingood, N.B., Hrycak, P., 1970. Survey of Literature on Flow Characteristics of a Single Turbulent Jet Impinging on a Flat Plate, NASA TN D-5652, NTIS N70-18963. NASA, Lewis Research Center, Washington, DC.

Goodyer, M.J., 1975. A New Probe for the Direct Measurement of Stagnation Pressure in Supersonic Flow, Aeronautical Research Council Reports and Memoranda. Her Majesty's Stationery Office, London.

Gracey, W., 1957. Wind-tunnel Investigation of a Number of Total-pressure Tubes at High Angles of Attacks, National Advisory Committee for Aeronautics Report 1303.

Hodson, H.P., 1996. Turbomachinery Aerodynamics-Experimental Methods, University of Cambridge Programme for Industry, Cambridge, UK.

Jambunathan, K., Lai, E., Moss, M.A., Button, B.L., 1992. A review of heat-transfer data for single circular jet impingement. Int. J. Heat Fluid Fl. 13 (2), 106–115.

Katz, J., Plotkin, A., 2004. Low-speed Aerodynamics, second ed. Cambridge Aerospace Series, Cambridhe, UK.

Kays, W.M., London, A.L., 1984. Compact Heat Exchangers. Krieger Publishing Company, Malabar, FL.

Martin, H., 1977. Heat and mass transfer between impinging gas jets and solid surfaces. Advances in Heat Transfer 13, 1–60.

McNaughton, K., Sinclair, C., 1966. Submerged jets in short cylindrical flow vessels. J. Fluid Mech. 25 (2), 367–375.

Park, C., 1984. Injection-induced turbulence in stagnation-point boundary layers. AIAA J. 22, 219–225.

R.E. Rayle (1959), Influence of Orifice Geometry on Static Pressure Measurements, ASME Paper 59-A-234.

Reungoat, D., Rivière, N., Fauré, J.-P., 2007. 3C PIV and PLIF measurement in turbulent mixing. J. Visual. – Japan 10 (1), 99–110.

Schlichting, H., 1979. Boundary-layer Theory. McGraw-Hill, Inc., New York.

Sparrow, E.M., Yanez Moreno, A.A.AT Effect of yaw on forced convection heat transfer from a circular cylinder, 1987. Int. J. Heat Mass Tran. 30, 427–436.

Thompson, J.S., Holder, D.W., 1958. Notes on Wind-tunnel Pressure Measurements from the Operator's Point of View, Royal Aircraft Establishment Tech. Note No. 2547.

Zdravkovich, M.M., 1997. Flow Around Circular Cylinders Vol. 1 Oxford Science Publications, Oxford.

Zdravkovich, M.M., 2003. Flow Around Circular Cylinders Vol. 2 Oxford Science Publications, Oxford.

CHAPTER 4

Fast-response Velocity and Shear Stress Measurements

D. Barratt*, M.A. Nguepnang, M.D. Atkins†**
*Department of Mechanical and Process Engineering, ETH Zürich, Zürich, Switzerland
**Digital Appliance, Africa Regional Headquarters, Samsung Electronics, Johannesburg, South Africa
†Atkins Technology, Johannesburg, South Africa

4.1 INTRODUCTION

All flows exhibit time variations in velocity, although the characteristics of the fluctuations differ. The shedding of laminar vortices from a cylinder in crossflow (Fig. 4.1) exhibits periodic fluctuations in velocity with a single, recurrent timescale (*tonal fluctuation*). An increase in Reynolds number will ultimately cause a transition to turbulence in the cylinder wake (Fig. 4.2), which introduces random velocity fluctuations with a broad range of timescales (*broadband fluctuation*).

A complete description of a flow field should thus include a description of the fluctuations in velocity with several sensors available for experimental characterization. Selection of a sensor depends primarily upon the "response time," that is, the time required for a change in velocity to be detected.

The turbulent wake shed from a circular cylinder in crossflow is depicted in Fig. 4.2, containing two significant timescales. The timescale of the vortex shedding (Δt_1) significantly exceeds the timescale of the turbulence velocity fluctuations (Δt_2). A sensor with a response time less than Δt_1 could be used to characterize the velocity fluctuations induced by vortex shedding but a sensor with a response time less than Δt_2 would be required to characterize the turbulence velocity fluctuations. The complexity and cost of a sensor typically increase substantially if a significantly smaller response time is required. In this chapter, three widely used unsteady sensors are addressed:

- Hot-wire anemometry for unsteady velocity measurement
- Hot-film sensor for shear stress measurement
- Particle image velocimetry (PIV) for velocity measurement in a plane of flow

Application of Thermo-Fluidic Measurement Techniques
http://dx.doi.org/10.1016/B978-0-12-809731-1.00004-6

Figure 4.1 ***Vortex shedding from a cylinder in crossflow in the laminar regime with unsteady velocity at point* P.**

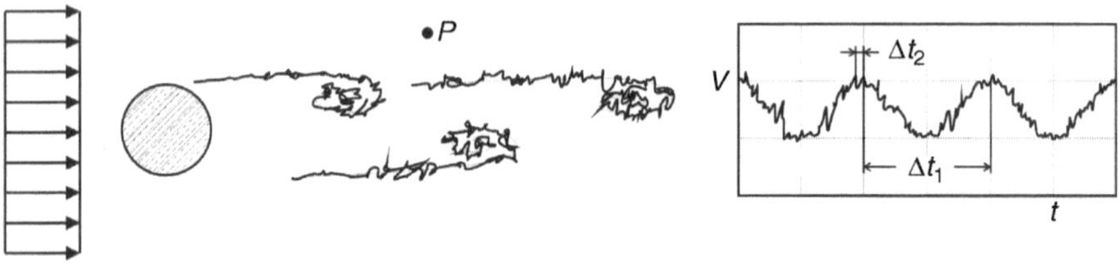

Figure 4.2 ***Vortex shedding from a cylinder in crossflow in the turbulent regime with unsteady velocity at point* P.**

4.2 HOT-WIRE ANEMOMETRY

4.2.1 What is it?

A hot-wire anemometer is a thermal transducer which has been widely used to measure instantaneous flow velocity. The use of the hot-wire anemometer permits instantaneous flow velocity to be calculated from electric voltage measurements. The advantage of hot-wire anemometry is associated with its very high spatial resolution and excellent frequency response characteristics. In most cases, the probe-stem is perpendicularly aligned with the mainstream direction of flow. A particular type of hot-wire is given in Fig. 4.3. For this configuration, the direction of mainstream flow coincides with the x-axis and the axis of the hot-wire probe is initially aligned with the y-axis.

The most essential part of the hot-wire anemometer is the thin wire sensor where forced convective heat transfer takes place from the wire to flow over the wire. A comprehensive review can be found in Perry (1982) and Lomas (1986). Thus, this book focuses on its detailed applications to fundamental fluid mechanics issues with a brief description of the typical calibration procedure.

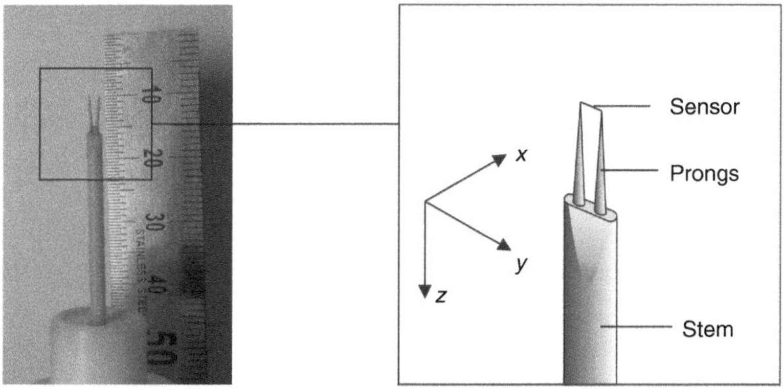

Figure 4.3 ***Configuration for coordinate system and hot-wire probe (Dantec Dynamics™ 55P11).***

4.2.2 How Does it Work?

When an electric current is passed through a fine wire exposed to crossflow, heat is transported from the heated wire to the flow. As the flow rate varies, the heat transfer from the wire varies accordingly. This, in turn, causes a variation in the heat balance of the wire. The wire is made from a material which possesses a comparatively high temperature coefficient of resistance. If the temperature of the wire varies, so does its resistance and hence the heating. A hot-wire anemometer is operated in constant temperature (CTA) mode. Since the resistance of a wire is proportional to its temperature, a constant resistance and hence constant temperature are maintained.

The physics of fluid flow and convective heat transfer are inextricably linked by relationships of the general form:

$$\mathrm{Nu} = f(\mathrm{Re}, \mathrm{Pr}, \mathrm{Kn}, \ldots \text{geometrical factors}) \tag{4.1}$$

where Nu is the Nusselt number, Re is the Reynolds number, Pr is the Prandtl number, and Kn is the Knudsen number. These are all nondimensional quantities.

With the assumption that only one flow parameter is varied, the heat loss can be interpreted as a measure of that parameter. In 1914, King derived a solution for the heat transfer from an infinitely long cylinder in an incompressible low Reynolds number flow, as:

$$\mathrm{Nu} = A + B(\mathrm{Re})^{n} \tag{4.2}$$

where A and B are dimensionless constants to be determined. If the flow properties and wire resistance remain constant, this expression reduces to:

$$(E)^2 = A + B(U)^n \tag{4.3}$$

where E is the electric voltage across the sensor and U is the flow velocity. Eq. (4.3) is called King's law. King suggested that $n = 0.5$, but a better curve fit was obtained by using $n = 0.45$ by Collis and Williams (1959).

4.2.3 Calibration

The process of calibration is undertaken to construct a relationship between the electric voltage (E) and the flow velocity (U) passing the hot-wire sensor in Eq. (4.3). In application, the electric voltage from the wire is read to calculate the flow velocity but in calibration a known flow velocity is set whilst the electric voltage from the wire is recorded to obtain the constants (A, B, and n) in Eq. (4.3).

During calibration, for example, the hot-wire anemometer is mounted on a calibration unit consisting of a rotary table, a free-jet generator, and a data acquisition system (Fig. 4.4). The free-jet generator is connected to a computer-controlled module. The jet velocity is controlled and monitored by a personal computer. The single-sensor probe consists of a long platinum-plated tungsten wire (diameter 5.0 μm and length 1.25 mm) supported by a probe stem (diameter 4.0 mm). The wire is welded directly to the prongs and the entire wire length acts as a sensor. The main purpose is to measure the mean and fluctuating velocities of 1D flow. The calibration is carried out by varying the velocity ranging from 5.0 to 30.0 m/s with a 2.5 m/s increment and from 30.0 to 60.0 m/s with a 5 m/s increment. In addition, the yaw angle is varied from −90 to +90° with a 5° increment to examine the effect of yaw angle on the measured velocity.

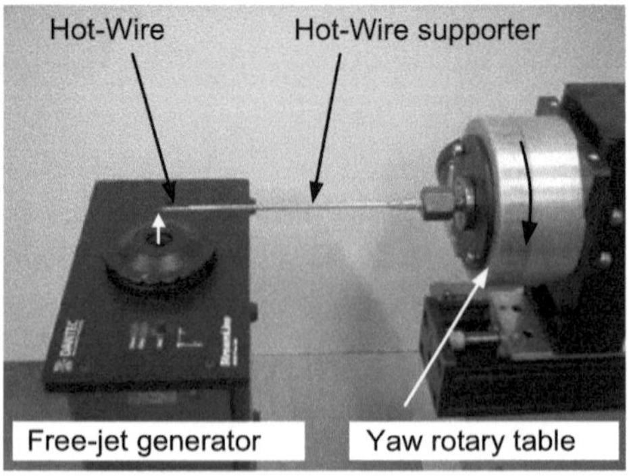

Figure 4.4 ***Photograph of calibration setup including a free-jet generator.***

4.2.4 Hot-Wire Calibration

4.2.4.1 Sensitivity of Electric Voltage to Flow Velocity

When flow velocity changes, wire temperature changes accordingly. In CTA mode, heat is generated to maintain the wire at constant temperature. As flow velocity is varied, the electric voltage exhibits a response as seen in Fig. 4.5a. The electric voltage increases to maintain the constant wire resistance as flow velocity increases. A concern of hot-wire anemometry is the error introduced by change in ambient temperature during a measurement. The calibration data shown in Fig. 4.5a have been taken before and after an experiment with a change in ambient temperature of 1.8 K. A maximum deviation of 0.6% occurs between the two calibration curves; a larger deviation can be anticipated for a more substantial temperature difference. The results of the calibration have been reprocessed with King's law (Eq. 4.3) in Fig. 4.5b, resulting in a linear curve fit with the coefficients $A = 1.133$ and $B = 1.925$ and a coefficient of determination (R^2) of 0.9978.

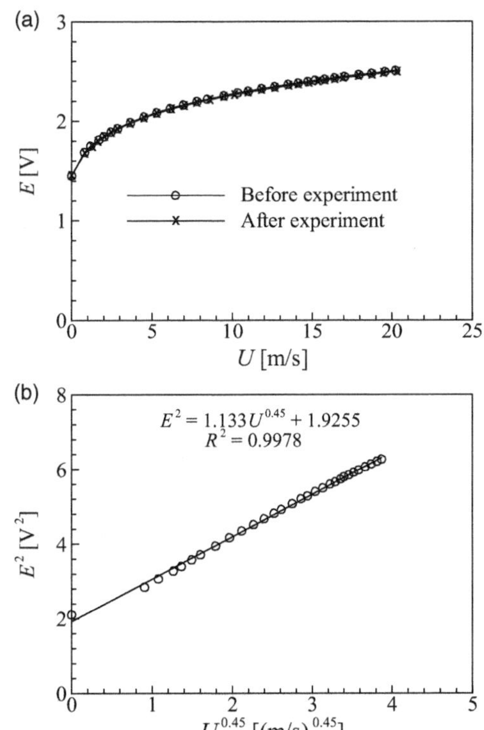

Figure 4.5 ***Sensitivity of electric voltage (E) to flow velocity (U).*** (a) *U* versus *E*; (b) $U^{0.45}$ versus E^2.

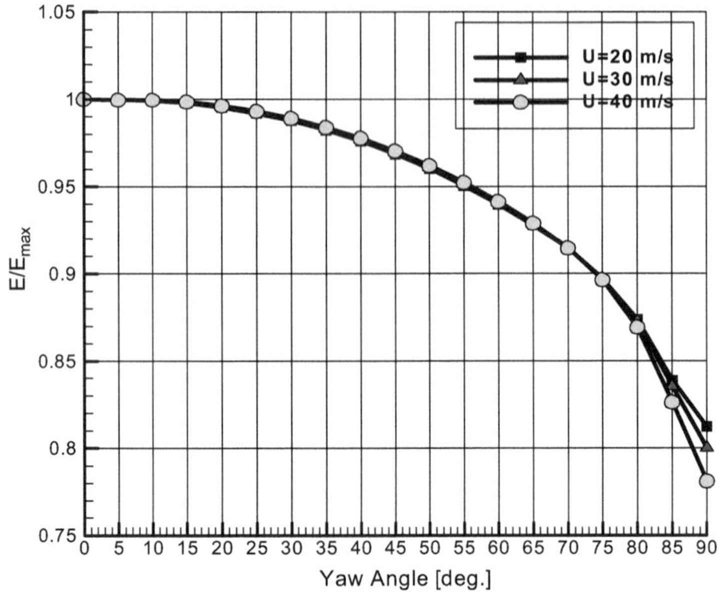

Figure 4.6 ***Effect of yaw angle.***

4.2.4.2 Effect of Yaw Angle

When the hot-wire anemometer is not placed perpendicular to the flow direction, there is a velocity component parallel to the axis of the wire which influences the cooling effect. Fig. 4.6 plots the hot-wire anemometer response as a function of yaw angle for selected flow velocities, with E representing the electric voltage across the wire. When the hot-wire anemometer is placed normal to the flow direction, the electric voltage is $E_{max.}$ In general, the electric voltage is symmetric with respect to the zero yaw angle due to the symmetry of the wire. The results of Fig. 4.6 demonstrate a monotonic decline in E with yaw angle. The effect of yaw shows Reynolds number insensitivity for yaw angles less than 75°. Larger yaw angles exhibit a dependence of flow velocity and thus a sensitivity to Reynolds number is observed. Ultimately, Fig. 4.6 indicates that small yaw angles (less than 10°) do not significantly affect the measured velocity.

Example 4.1 Wake Shedding Behind a Cylinder

A. Background

It is common to witness the formation of a Kármán eddy street when a fluid flows past a stationary cylinder placed normal to a mainstream flow. Fig. 4.7 shows an instantaneous image of periodic vortex shedding from an upright cylinder midspan at $Re_D = 1.0 \times 10^4$, which is obtained by ink-dye injection in a water tunnel where D is the cylinder diameter. Such a wake pattern may

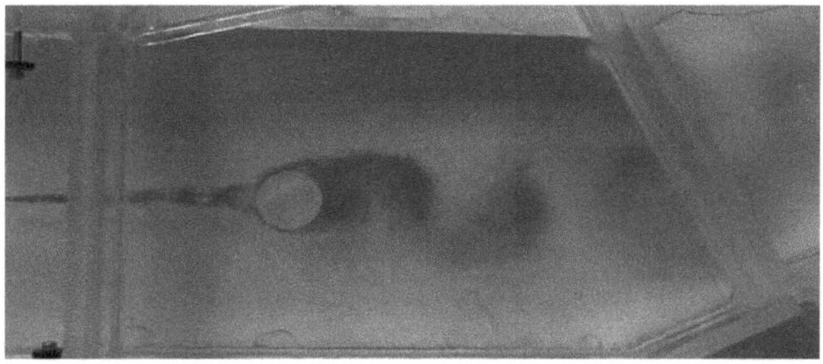

Figure 4.7 ***Wake shedding around a single cylinder visualized by ink injection.***

be quantitatively characterized with a dimensionless wake shedding frequency, that is, the Strouhal number ($S_t = f D / U$), where f is the shedding frequency.

Numerous studies have reported that the dimensionless wake frequency behind the 2D circular cylinder (ie, infinitely long) has a value of about 0.2 over a wide range of Reynolds numbers (Zdravkovich, 1997). In this example, the quantification of such unsteady wakes shed from a single cylinder is conducted by using hot-wire anemometry in a wind tunnel.

B. Experimental Details

The wake shedding frequency was measured at $0.5D$ downstream from one edge of the cylinder as illustrated in Fig. 4.8. A single-sensor hot-wire from

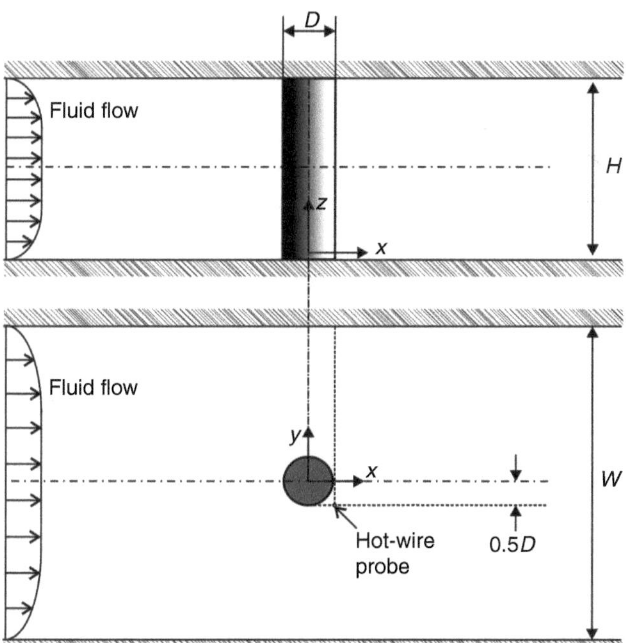

Figure 4.8 ***Wake shedding frequency measurement behind a cylinder using hot-wire anemometry (top: side view, bottom: plan view).***

Dantec™ was used. Prior to the experiments, the mainstream velocity was measured using a Pitot tube placed 3*D* upstream of the cylinder and it was fixed at 10 m/s during the entire experiment with a fixed cylinder diameter of 30.0 mm. Fig. 4.9a and b show a hot-wire probe placed downstream of the cylinder in a wind tunnel.

C. Discussion of Results

Fig. 4.10 presents the raw signal data from the hot-wire probe. The signal fluctuates in time, showing a dominant period every 15.6 ms. Therefore, the wake shedding frequency was calculated to be approximately 64.1 Hz (= 1/0.0156 s). In the present example, the dimensionless wake frequency (S_t) was found to be 0.192 at $Re_D = 1.0 \times 10^4$, showing good agreement with data reported in the literature (Zdravkovich, 1997).

(a)

(b)

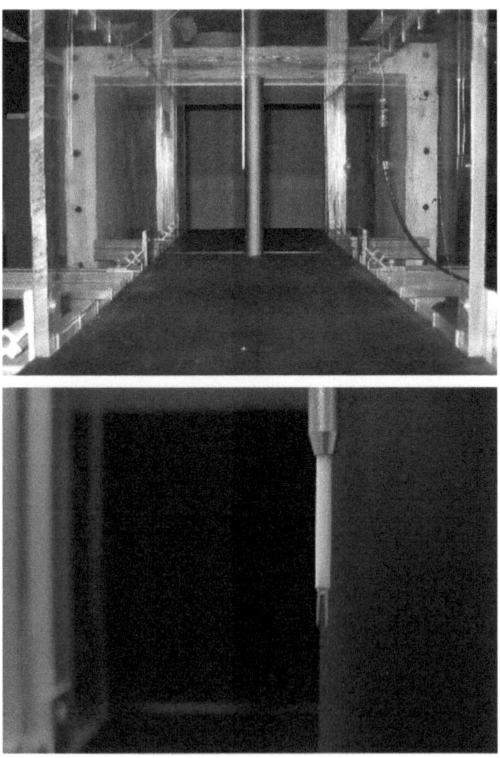

Figure 4.9 ***Test setup. (a) Test section with a single cylinder; (b) hot-wire probe located immediately downstream of the cylinder.***

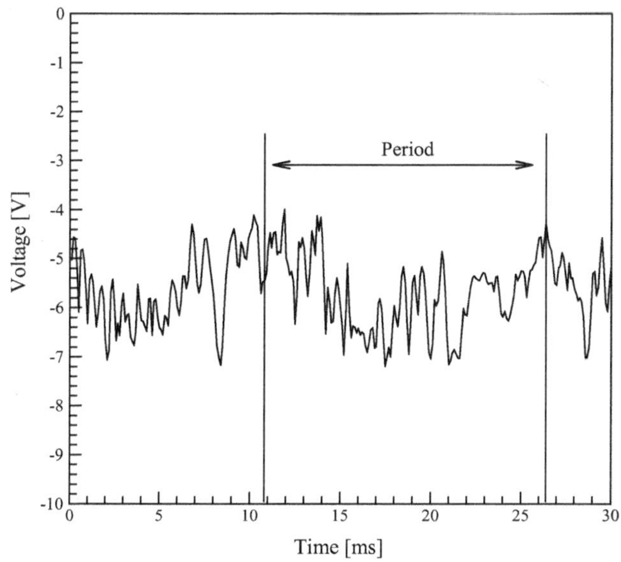

Figure 4.10 ***Raw data signal from the hot-wire probe at the set mean flow velocity of 10 m/s.***

Example 4.2 Turbulence in Free-Jets

A. Background

The impingement of a round jet on a flat surface is a commonly encountered configuration in engineering applications. Heat transfer at the stagnation point is of concern as an indicator of the overall heat transfer as well as a source of thermal stress. Previous studies have shown that stagnation point heat transfer varies with the impingement distance (z) of the jet (Martin, 1977) and peak stagnation point heat transfer occurs at an impingement distance of roughly seven jet diameters (D_j). The present example provides an explanation based on the structure of a "free-jet."

A jet ejected out of a nozzle without any downstream obstruction is termed a free-jet. The characteristics of a free-jet show sensitivity to Reynolds number (Re_j), primarily due to the onset of turbulence. Conventionally, the Reynolds number of a free-jet is defined with the diameter of the jet (D_j) as the characteristic length and the centerline velocity at the jet exit (w_e) as the characteristic velocity:

$$Re_j = \frac{\rho w_e D_j}{\mu} \tag{4.4}$$

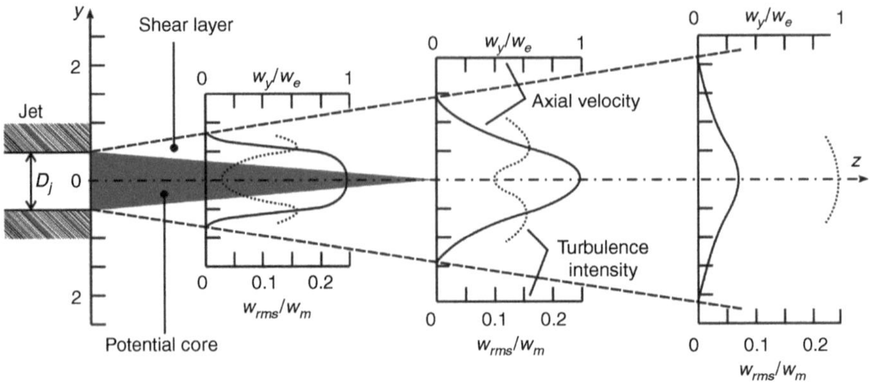

Figure 4.11 ***Schematic of axial velocity and turbulence in the jet (Gardon and Akfirat, 1965).***

where ρ is the density of the fluid and μ is the kinematic viscosity of the fluid. A sufficiently high Reynolds number results in the formation of a fully turbulent jet which forms the basis of most engineering applications, with an extensive review by Abramovich (1963).

The structure of a fully turbulent free-jet is schematically shown in Fig. 4.11 (after Gardon and Akfirat, 1965). As the flow leaves the nozzle, shear layers develop at the edge of the jet. Instability of the shear layers results in a mixing process which causes the shear layers to spread radially inwards until the shear layers eventually merge at the center of the jet. The core of the jet that remains unaffected by the merger process is known as the "potential core." The level of turbulence inside and outside the potential core can be quantified with the root-mean-square (RMS) of the streamwise velocity fluctuations:

$$w_{rms} = \sqrt{\frac{1}{N}\sum_{1}^{N}(w_i')^2} \tag{4.5}$$

Normalization of w_{rms} with the centerline velocity at the jet exit (w_e) provides an indicator of the strength of the turbulence within the jet (Fig. 4.11). Inside the potential core and during the merger process, peak turbulence is observed in the center of the shear layer. After the merger process has completed, peak turbulence is observed at the center of the jet. This example considers the level of turbulence along the centerline of the jet, both inside and outside the potential core. The turbulence distribution along the centerline has been used to explain the peak in stagnation point heat transfer observed at an impingement distance of $7Dj$.

B. Experimental Details

To quantify the level of turbulence along the centerline of the round jet, the time-resolved velocity along the jet axis has been measured using a

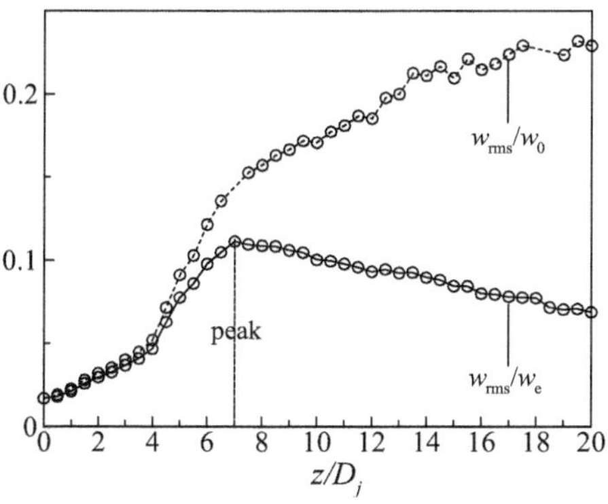

Figure 4.12 ***Variation of turbulence strength (w_{rms}/w_e) along the axis of a round jet at $Re_{DJ} = 20{,}000$.***

single-sensor hot-wire probe (Dantec Dynamics MiniCTA™) calibrated prior to the experiment. The measurements have been performed at a sampling frequency of 20 kHz over a duration of 10 s. The experimental uncertainty of w_{rms}/w_e has been estimated to be within 4.2% based on a method reported in Coleman and Steele (1999).

C. Discussion of Results

The variation of turbulence strength (w_{rms}/w_e) along the centerline of the jet has been plotted in Fig. 4.12. The peak in turbulence strength observed at $z/D_j = 7.0$ is considered responsible for the peak in stagnation point heat transfer for a flat plate located at the same impingement distance. As reference, there is a conventional parameter, termed "turbulence intensity (w_{rms}/w_0)," which characterizes the level of turbulence at a point. No peak in turbulence intensity exists along the jet axis.

4.3 HOT-FILM ANEMOMETRY

4.3.1 What is it?

The fragility of hot-wire probes combined with the difficulty in measurement access has motivated the advent of hot-film probes (Fig. 4.13a) which are robust and suitable for flush surface-mounting (Fig. 4.13b). The hot-film probe is comprised of metal-film sensors which are deposited on a quartz or polymer substrate for electrical insulation with a further coating of quartz or lacquer to prevent abrasion and fouling of the exposed sensor surfaces. The high breakdown voltage of quartz ensures suitability across a broad range of temperatures while the thermal coefficient of expansion is compatible with the commonly used nickel sensors, thus preventing distortion. Despite the mechanical and electrical importance of coating the sensors, the

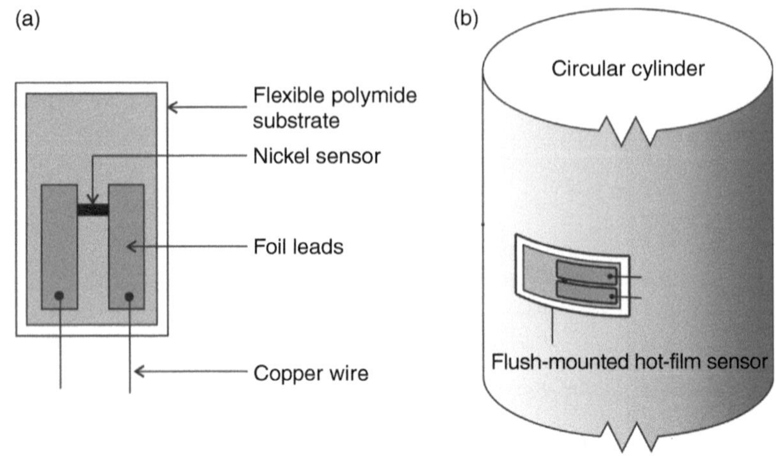

Figure 4.13 ***Hot-film sensor.*** (a) Components (Cook, 1994); (b) flush-mounted configuration.

coatings should be kept as thin as possible to minimize the thermal capacity of the probe as it limits the frequency response (Lomas, 1986).

4.3.2 How Does it Work?

Analogous to the hot-wire probe, the hot-film probe forms part of a Wheatstone bridge which is balanced to maintain the temperature and resistance of the probe by adjusting the current through the circuit [Constant Temperature Anemometry (CTA)]. A theoretical framework allows the wall shear experienced by the sensor to be related to the differential voltage across the sensor.

Fig. 4.14 portrays heat transfer from the hot-film (Q_j) which occurs through convection into the overpassing flow (Q_f) and conduction into the surface substrate (Q_s), yielding the relation:

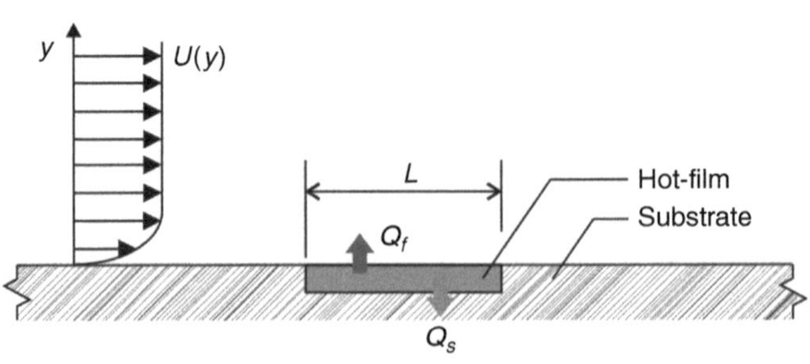

Figure 4.14 ***Flush-mounted hot-film sensor.***

$$Q_j = Q_f + Q_s \tag{4.6}$$

The convective heat transfer from the film is governed by Newton's Law of Cooling:

$$Q_f = hwL\left(T_\infty - T_w\right) \tag{4.7}$$

which depends on the convection coefficient (h), the length (L) and width (w) of the sensor, and the temperature differential between the sensor (T_w) and overpassing flow (T_∞). Combination of Newton's Law of Cooling with the definition of the Nusselt number:

$$Nu_L = \frac{hL}{k} \tag{4.8}$$

yields the linear relation:

$$Q_f = C_1 Nu_L \tag{4.9}$$

where the proportionality constant (C_1) depends on the width of the sensor, the thermal conductivity of the fluid (k), and the temperature differential between the sensor and the free-stream which is fixed by the CTA system. Based on the result of Ludwieg (1950), it can be shown that the Nusselt number for a hot-film sensor is given by:

$$Nu_L = 0.807\left(\mathrm{Re}_L\right)^{2/3}\mathrm{Pr}^{1/3} \tag{4.10}$$

where the characteristic velocity of the Reynolds number (Re) is the friction velocity ($u^\star$):

$$u^\star = \sqrt{\frac{\tau_w}{\rho}} \tag{4.11}$$

which is comprised of the local wall shear (τ_w) and fluid density (ρ).

The heat transfer relation for a hot-film thus deviates from King's law (Eq. 4.3) applied to hot-wire sensors. Combination of the friction velocity (Eq. 4.11) with the Nusselt number relation (Eq. 4.10) and the expression for convection heat transfer (Eq. 4.9) allows the energy balance (Eq. 4.6) to be expressed as a function of wall shear. In CTA mode, the resistance of the sensor is constant and the total heat produced by the sensor (Q_j) is proportional to the square of the voltage across the sensor (E^2), reducing the energy balance (Eq. 4.6) to the relation:

$$E^2 = A\tau_w^{1/3} + B \tag{4.12}$$

which provides a semiempirical basis for hot-film calibration through the experimental determination of constants A and B.

4.3.2 Hot-Film Calibration

Calibration of a hot-film can be accomplished by immersing the flush-mounted sensor in a free-stream. The voltage across the sensor (E) should be recorded together with the wall shear (τ_w) acting on the sensor which may be calculated from a velocity profile measured with hot-wire anemometry. The definition of wall shear:

$$\tau_w = \mu \left.\frac{du}{dy}\right|_{y=0} \tag{4.13}$$

depends upon the dynamic viscosity of the fluid (μ) and velocity gradient normal to the wall at the surface, and can be determined from a linear fit of the velocity data nearest to the wall, exemplified in Fig. 4.15.

Determination of the sensor voltage and wall shear for a range of free-stream velocities allows the sensor voltage to be correlated with wall shear to determine the calibration constants A and B (Eq. 4.12) by linear regression, as depicted in Fig. 4.16.

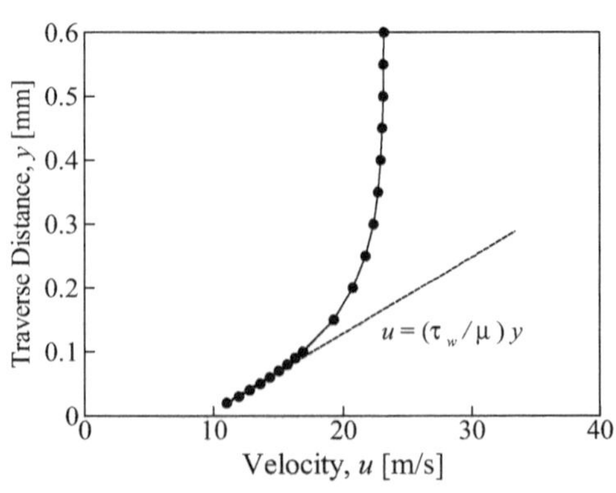

Figure 4.15 ***Wall shear calculated from measured velocity profiles.***

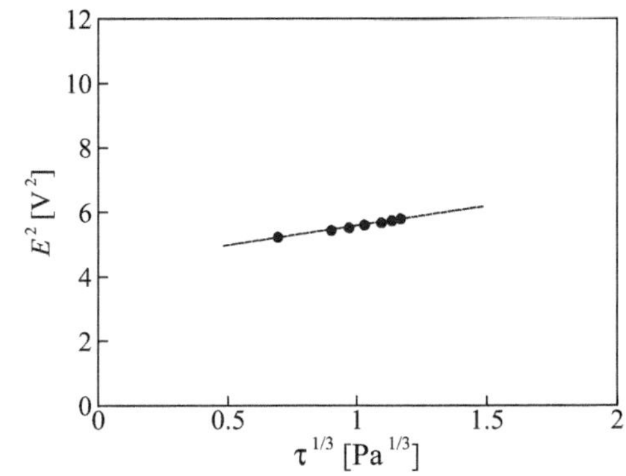

Figure 4.16 ***Experimental determination of calibration constants for a hot-film.***

Example 4.3 Round Jet Impingement on a Circular Cylinder Inside the Potential Core

A. Background

The impingement of a single round jet on a circular cylinder (Fig. 4.17a) is a fundamental configuration with relevance to numerous engineering applications, including the internal cooling of a turbine blade, furnace cooling, and the cooling of extruded materials. Flow impinging upon the cylinder initially stagnates at the leading edge and thereafter accelerates away from the stagnation point, resulting in the development of a laminar boundary layer which can be expected to transition if the Reynolds number of the jet is sufficiently high (Fig. 4.17b). Transition influences the heat transfer characteristics of the configuration and the occurrence of transition may be investigated with a flush-mounted hot-film sensor on the surface of the cylinder.

B. Experimental Details

The test facility consists of a test bench (Fig. 4.18) which receives air supply from a wind tunnel. A plenum chamber with flow straighteners breaks apart large-scale coherent eddies followed by a sudden contraction intended to improve flow uniformity at the nozzle exit (diameter 20 mm). A single hot-film sensor (55R47 Dantec Inc.) of thickness 50 μm has been flush mounted on the target cylinder, diameter 50 mm, depicted in Fig. 4.18. The cylinder has been mounted on a rotary table (Velmex B4800TS) to allow for angular rotation in the range $0° \leq \alpha \leq 120°$ at 5° intervals with an uncertainty of ±0.2°.

A diameter ratio (D/D_j) of 2.5 has been used with an impingement distance (z/D_j) of 2. The Reynolds number incorporates the diameter of the cylinder (D) as the characteristic length and the centerline velocity at the jet exit (w_e) as the characteristic velocity:

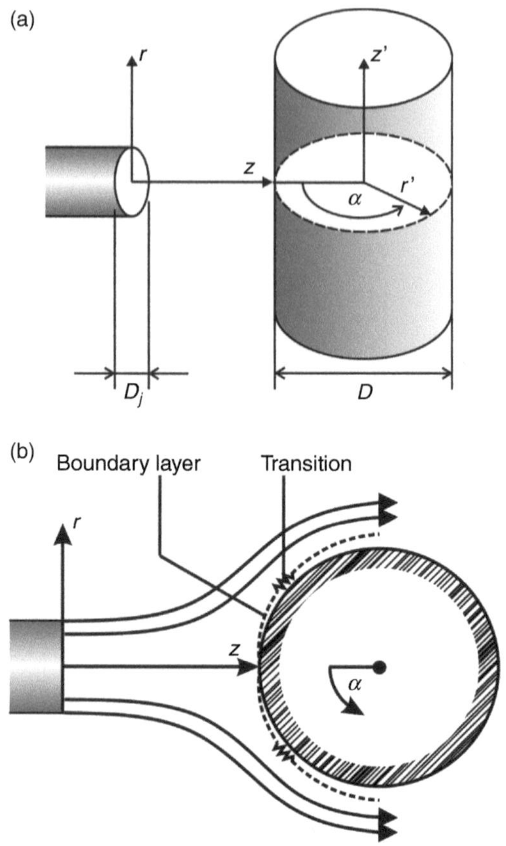

Figure 4.17 ***Round-jet impingement on a circular cylinder.*** (a) Configuration; (b) flow field depicting transition of the boundary layer.

$$\mathrm{Re}_D = \frac{\rho w_e D}{\mu} \tag{4.14}$$

The density and dynamic viscosity of air are denoted by ρ and μ, respectively. The Reynolds (Re_D) number has been fixed at 50,000.

C. Discussion of Results

The shear stress (τ_w) distribution in the crossflow plane has been normalized to determine the skin friction coefficient (C_f):

$$C_f = \frac{\tau_w}{\rho w_e^2/2} \tag{4.15}$$

The wall-shear distribution in the crossflow plane is shown in Fig. 4.19 with the skin friction coefficient (C_f) on the ordinate and the

Figure 4.18 ***Test bench setup.***

azimuth angle (α) on the abscissa. The bars indicate the RMS of the measured wall-shear fluctuations at each location. Stagnation results in zero shear at the leading edge of the cylinder ($\alpha = 0°$) followed by flow acceleration away from the stagnation point which causes shear to increase until 25°, before the effect of boundary layer growth becomes dominant and shear starts to decline. The decline in shear is interrupted by a plateau which coincides with a sudden increase in the wall-shear fluctuations

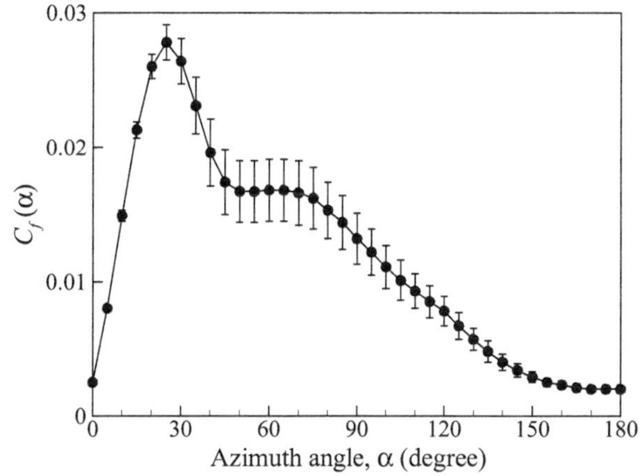

Figure 4.19 ***Wall-stress distribution in the crossflow plane with bars indicating fluctuation.***

indicative of the velocity fluctuations induced by transition. The velocity fluctuations convect momentum towards the wall, counteracting the diffusive effect of boundary layer growth and thus causing the fluctuation and plateau of wall shear. Following transition, a monotonic decline in shear occurs on the aft of the cylinder as a result of continued growth of the boundary layer. Shear stress measurements with the hot-film sensor thus provide a qualitative indication of transition on the surface of the cylinder.

4.4 PARTICLE IMAGE VELOCIMETRY

Hot-wire and hot-film share the same weaknesses of being intrusive to the flow and limited to point measurements in space. In contrast, particle image velocimetry (PIV) is an indirect optical measurement technique which can provide measurements across a plane of flow. Measurement of the velocity field is accomplished with raw digital images that are compared in successive frames separated by a known time interval to provide an instantaneous representation of the velocity field.

A limitation of PIV measurements is the typical maximum recording frequency of ≈ 15.0 Hz which stems from the data storage and transfer speed of the digital CCD camera and the maximum repetition rate of the pulsed laser system. Consequently, the temporal resolution of a PIV system is often insufficient to resolve unsteady flow features. However, an ensemble of instantaneous PIV velocity field measurements may still yield quantitative information about the unsteadiness of the flow.

An inherent assumption of PIV measurement is that the trajectory of the seeding particles is the same as the actual flow field and the particles do not interact with one another (Westerweel, 1997), which is considered reasonable in low-speed flows (Adrian, 1991). It is further assumed that the measured flow field is predominately 2D so that the majority of particles entering into the light sheet remain within the light sheet during the measurement period.

Example 4.4 Flow Behind two Side-by-Side Cylinders

A. Background

The wake behind side-by-side circular cylinders has been shown by Sumner et al. (1999) to be influenced by the dimensionless spacing between the cylinders (T/D) where T is the center-to-center distance and D is the cylinder diameter. The flow between the cylinders, termed the "gap flow," exhibits biased bistable behavior characterized by deflection towards the upper or

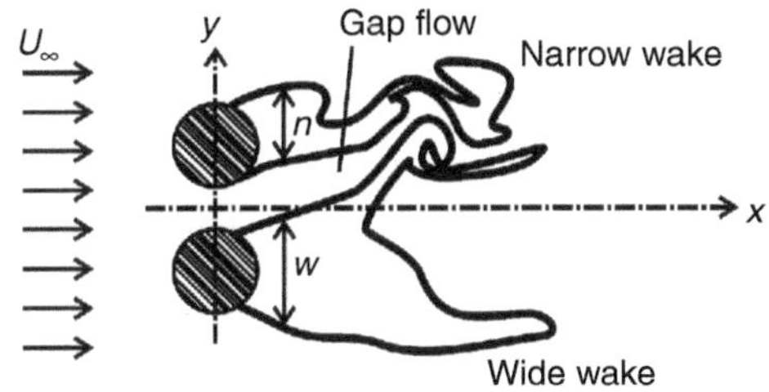

Figure 4.20 ***Schematic of the wake pattern formed in the biased flow regime with two side-by-side circular cylinders 1.1 < T/D < 2.2.***

lower cylinder with switching at random intervals (Fig. 4.20). The asymmetric wake structure, with respect to the central axis (x-axis), arises in the range $1.1 < T/D < 2.2$ with a symmetric wake structure developing for $T/D > 2.2$.

The mechanism of the biased and bistable gap flow remains unclear; however, speculative arguments have been put forth by Wang and Zhou (2005) indicating that the behavior originates from shear layer interaction between the gap flow and the near-wake region. A demonstration follows in which PIV is used to identify the shear layers and wake structure downstream of side-by-side cylinders.

B. Experimental Details

B1. Test Rig and Instrumentation

The experimental setup is depicted in Fig. 4.21 which shows a section of the open suction-type wind tunnel including two horizontally mounted

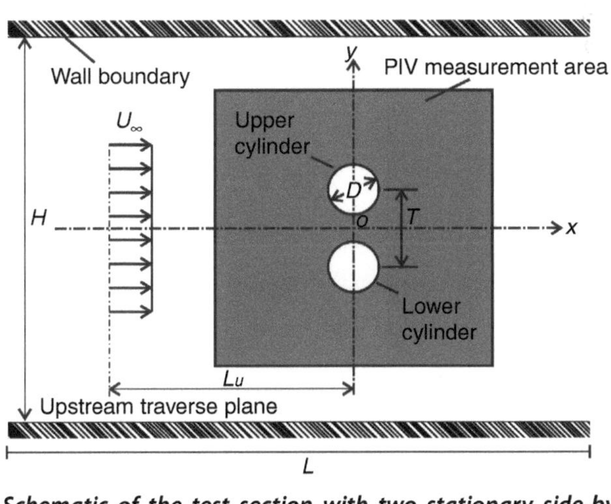

Figure 4.21 ***Schematic of the test section with two stationary side-by-side circular cylinders (side view).***

circular cylinders (diameter 15.0 mm and aspect ratio 20:8) with a Reynolds number of $Re_D = 6000$ based on the cylinder diameter.

B2. Data Acquisition

The upstream and downstream velocity fields, including the gap region, have been mapped simultaneously with particle image velocimetry (PIV) using two cylinders made of transparent glass tube and specialist laser optics (Atkins and Kim, 2015) placed inside the upper transparent cylinder (Fig. 4.22a). A 90° cone mirror transforms a laser beam into a "laser light disc" with 360° illumination around the upper cylinder in a plane perpendicular to the cylinder axis and located at the midspan (Fig. 4.22b). The flow field has been seeded with atomized mineral oil with a mean particle diameter of 1.0 μm and a laser light sheet of thickness 1.0 mm produced by a Nd:YAG laser (New Wave Research™) with a light wavelength of 532 nm and a pulse interval of 33 μs. The optical axis of the camera has been arranged perpendicular to the laser light sheet and focused at the midspan of the cylinders.

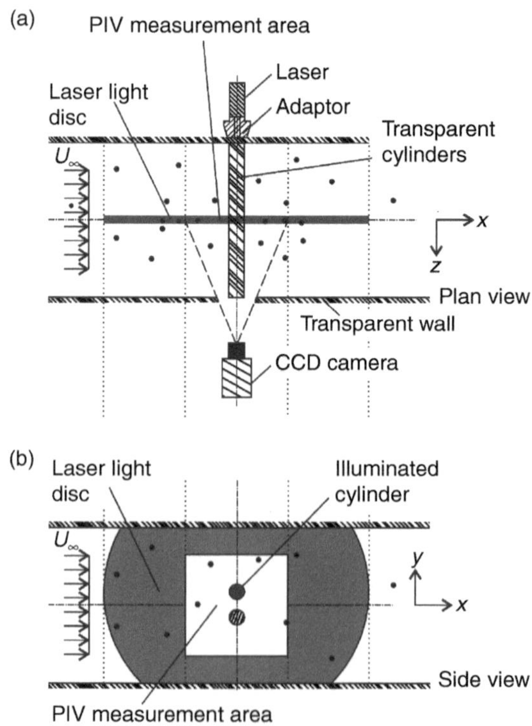

Figure 4.22 ***Experimental setup of the new PIV laser illumination configuration.*** Internal laser illumination method (a) plan view; (b) side view.

The flow field has been recorded with particle images that are mapped onto the CCD sensor with a pixel resolution of 2048 × 2048 and a pixel pitch of 7.4 μm. Division of the image plane into overlapping interrogation areas of 16 × 16 pixels allows the displacement field to be calculated with the Dantec™ DynamicStudio V1.45 software. The resultant vector spacing of 0.7 mm produces a 255 × 255 vector map converged iteratively by variation of the interrogation window offset (Westerweel et al., 1997). Removal of outliers has been achieved with peak detection validation, the exclusion of vectors which exceed a predefined value (ie, 12.0 m/s), and the incorporation of moving averaging. Less than 1% of the vectors were rejected after elimination of poorly illuminated regions and regions obstructed by the cylinders.

The time-averaged velocity field has been evaluated with an ensemble of 80 instantaneous vector maps sampled at a frequency of 5.0 Hz which allows for calculation of the RMS velocity map derived from decomposition of the instantaneous velocity (V) into a time-averaged ($\overline{V}$) and a fluctuating component (V'):

$$V = \overline{V} + V' \tag{4.16}$$

followed by normalization of the RMS velocity fluctuation with the free-stream velocity (U_∞):

$$V_{rms} = \frac{\sqrt{\left(V - \overline{V}\right)^2}}{U_\infty} \tag{4.17}$$

yielding the results shown in Fig. 4.23a and b.

C. Discussion of Results

Unsteady regions of flow for the two spacing ratios T/D = 1.1 and T/D = 2.1 can be identified with the RMS velocity map shown in Fig. 4.23. For closely spaced cylinders (T/D = 1.1), Fig. 4.23a shows narrow bands of increased velocity fluctuation on the outer surfaces of the cylinders, indicating the presence of shear layers where the free-stream fluid mixes with the wake. In the region directly behind the cylinders, asymmetry in the RMS velocity map indicates deflection of the gap flow towards the upper cylinder during the measurement period. Fig. 4.23b shows that a larger spacing between the cylinders (T/D = 2.1) causes two distinct wakes to emerge behind the cylinders with penetration of the gap flow into the wake region. Two bands of increased velocity fluctuation have also appeared near the inner cylinder surfaces, indicating the formation of shear layers where the gap flow mixes with the cylinder wakes. The flow field has now become nearly symmetric with respect to the central axis (x-axis).

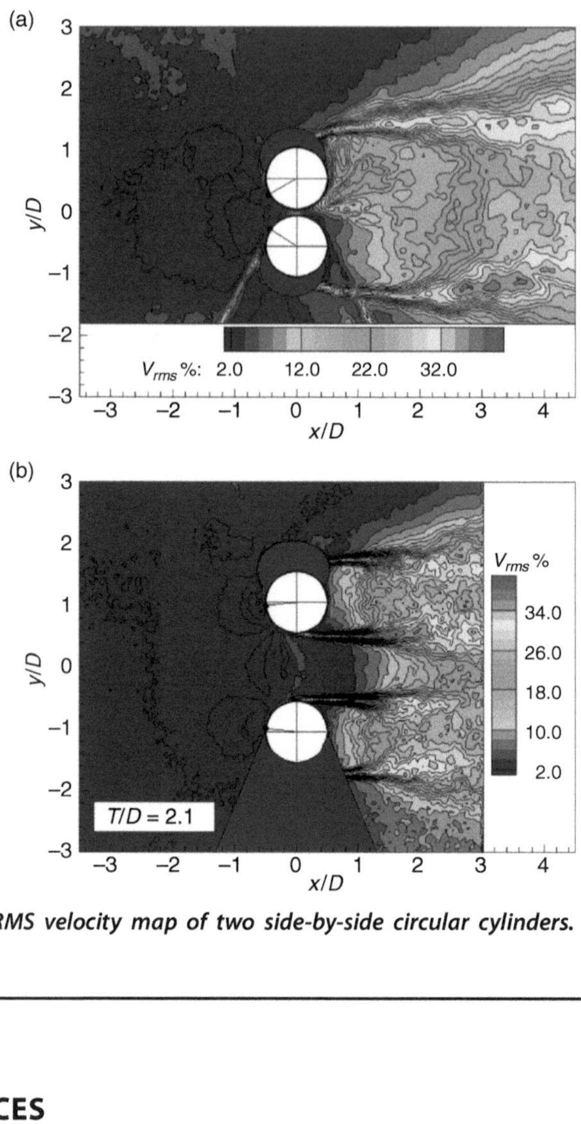

Figure 4.23 ***RMS velocity map of two side-by-side circular cylinders.*** (a) $T/D = 1.1$; (b) $T/D = 2.1$.

REFERENCES

Abramovich, G.N., 1963. The Theory of Turbulent Jets. MIT Press, Cambridge, MA, USA, pp. 1-5.

Adrian, R.J., 1991. Particle-imaging techniques for experimental fluid mechanics. Annu. Rev. Fluid Mech. 23, 261–304.

Atkins, M.D., Kim, T., 2015. Isotropic-planar illumination for PIV experiments. Exp. Fluids 56, 63.

Coleman, H.W., Steele, W.G., 1999. Experimentation and Uncertainty Analysis for Engineers, second ed. Wiley-Interscience, John Wiley & Sons, Inc., New York, NY, USA.

Collis, D.C., Williams, M.J., 1959. Two-dimensional convection from heating wires at low Reynolds numbers. J. Fluid Mech. 6, 357–389.

Cook, W.J., 1994. Response of hot-element wall shear stress gages in unsteady turbulent flows. AIAA J. 32, 1464–1471.
Gardon, R., Akfirat, J.C., 1965. The role of turbulence in determining the heat-transfer characteristics of impinging jets. Int. J. Heat Mass Tran. 8 (10), 1261–1272.
Lomas, C.G., 1986. Fundamentals of Hot Wire Anemometry. Cambridge University Press, Cambridge, UK.
Ludwieg, H., 1950. Instrument for Measuring the Wall Shearing Stress of Turbulent Boundary Layers, NACA Technical Memorandum 1284.
Martin, H., 1977. Heat and mass transfer between impinging gas jets and solid surfaces. Advances in Heat Transfer 13, 1–60.
Perry, A.E., 1982. Hot-wire Anemometry. Oxford University Press, Oxford.
Sumner, D., Wong, S.S.T., Price, S.J., Paidoussis, M.P., 1999. Fluid behavior of side-by side circular cylinders in steady cross-flow. J. Fluid. Struct. 13, 309–338.
Wang, Z.J., Zhou, Y., 2005. Vortex interaction in a two side-by-side cylinder near-wake. Int, J. Heat Fluid Fl. 26, 362–377.
Westerweel, J., 1997. Fundamentals of digital particle image velocimetry. Meas. Sci. Technol. 8, 1379–1392.
Westerweel, J., Dabiri, D., Gharib, M., 1997. The effect of discrete window offset on the accuracy of cross-correlation analysis of digital PIV recordings. Exp. Fluids 23, 20–28.
Zdravkovich, M.M., 1997. Flow Around Circular CylindersVol. 1Oxford Science Publications, Oxford.

CHAPTER 5

Velocity Field Measurement Using Particle Image Velocimetry (PIV)

M.D. Atkins
Atkins Technology, Johannesburg, South Africa

5.1 INTRODUCTION

Fluid flow velocity can be measured through various methods that are broadly classified as either *direct* or *indirect* (Raffel et al., 1998). Direct methods typically involve probes that are placed directly in the fluid flow and measure the velocity at a single point; examples of such methods include the Pitot tube and the hot-wire probe. Indirect methods, on the other hand, do not disturb the fluid flow and include particle image velocimetry (PIV) and laser Doppler anemometry (LDA).

PIV is an optical measurement technique which allows one to measure the velocity field by recording the transportation of *seeding* or *tracer* particles carried by the moving fluid. These seeding or tracer particles are typically introduced into the fluid flow as part of the experimental setup. The determination of the velocity field relies on recording images of the random pattern of tracer particles in the flow at two instances of time using a charge coupled device (CCD) digital camera. Ideally, the tracer particles should be small and neutrally buoyant so the trajectory of the particles closely follows the path of the fluid flow. The velocity is then calculated based on determining the displacement of the seeding particles recorded on two successive *frames* that are separated by the time interval (Δt) as shown in Fig. 5.1. Ultimately, the velocity field is extracted from digital images that were taken of a certain region of the flow field.

A key feature of PIV is that it is a quantitative method of flow field mapping where numerous instantaneous velocity vectors within the measurement region can be determined *simultaneously*. In contrast, to map the velocity field using point measurement techniques would require that the probe is traversed across the measurement region, which is time consuming and there is a possibility that a degree of physical insight into the flow behavior in question may not be obtainable. Furthermore, PIV is also regarded as a *nonintrusive* measurement technique which has the benefit of

Application of Thermo-Fluidic Measurement Techniques
http://dx.doi.org/10.1016/B978-0-12-809731-1.00005-8

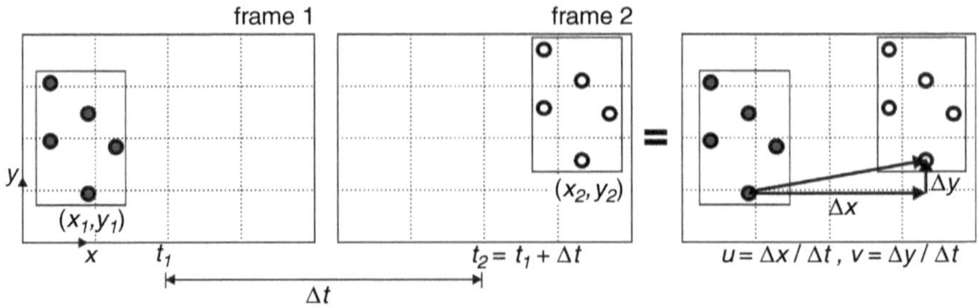

Figure 5.1 ***Illustration of how velocity components in the x–y plane are determined by a PIV system from a pair of successive frames at t_1 and $t_2 = t_1 + \Delta t$.***

not disturbing the flow field in the same manner as point measurement techniques, where probes are physically present in the flow field.

5.2 HOW PIV WORKS

A major challenge in the implementation of PIV is the determination of the *average* displacement vectors of local groups of seeding particles recorded on the frames (Keane and Adrian, 1992). PIV employs statistical analysis methods that require a *certain* concentration of seeding particles in the fluid flow to evaluate the velocity field. Over the last two decades a number of technological advances have changed the procedure of calculating the average particle displacements from a process that is manually intensive to one that is automated. Consequently, the PIV technique has reached a level of maturity where it is a commercially viable system for conducting accurate and reliable flow measurements.

A typical PIV system consists of five subsystems:

1. Flow test rig (such as a wind tunnel),
2. Seeding generator (used typically for seeding gaseous flows such as air),
3. Pulsed laser system with light sheet optics,
4. CCD digital camera system, and
5. Data acquisition and processing software.

In PIV experiments, seeding particles are supplied into the flow field using a seeding generator. Then, PIV measurements are based on the digital recordings of the random pattern of these seeding particles that are moving with the flow. The velocity field is determined within a measurement area illustrated in Fig. 5.2. It is imperative that the recordings are of a high quality as they will facilitate further analysis using the data acquisition and processing software.

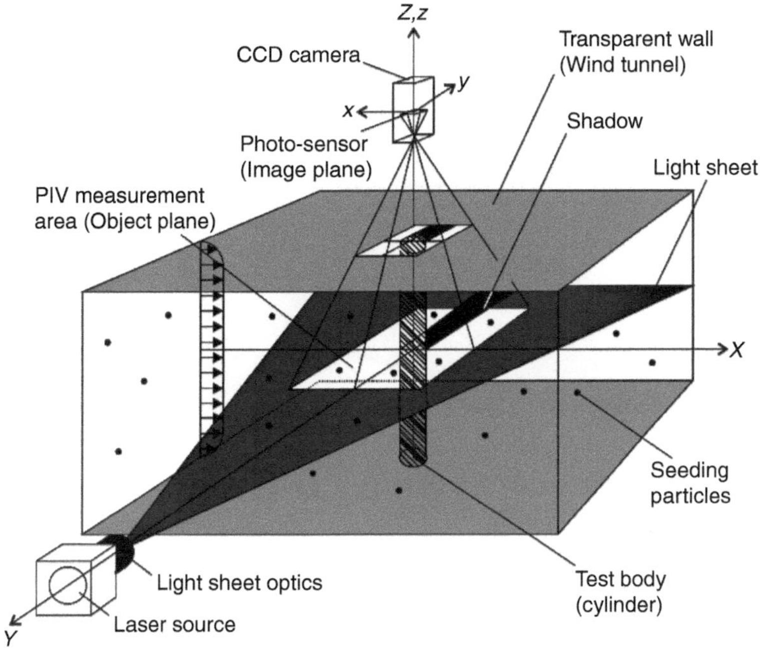

Figure 5.2 ***Typical 2D PIV setup with an external light sheet and a shadow behind a test body.***

The PIV recordings are created by synchronizing high-energy laser light sheet pulses with the exposure of the CCD camera. In planar or 2D PIV applications, laser light sheet optics transform the beam emitted from the laser into a thin laser light sheet. The seeding particles moving with the flow are illuminated when they fall within the plane of the light sheet. These seeding particles scatter the laser light that is then captured by a single CCD camera. The optical axis of the camera is perpendicular to the light sheet and the camera is focused so the object plane is coincident with the light sheet. The short-duration pulsed laser illumination ($\approx$ 10 ns) has the effect of "freezing" the particle motion on the PIV recordings so that the digital particle images are not blurred.

Scattered incident laser light is transmitted through the camera lens and received by a photosensitive sensor within the CCD camera. The photosensitive sensor is typically a square or rectangular array consisting of a number of photosensitive elements referred to as pixels (Willert and Gharib, 1991; Westerweel, 1997; Raffel et al., 1998; Adrian and Westerweel, 2011). The pixels create an analog signal by converting the incident light (photons) to electrical charges. The signal is then *quantized* into a digital output. In the analog to digital conversion process, a discrete pixel *gray value,* which

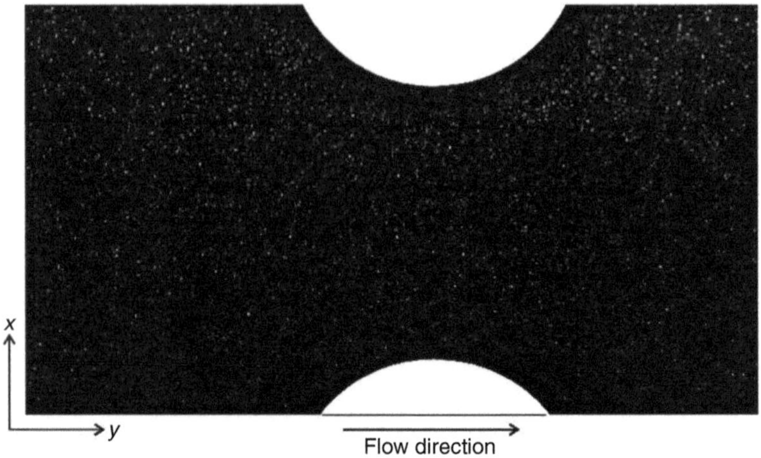

Figure 5.3 ***Raw PIV image of the fluid flow carrying seeding particles between two side-by-side circular cylinders.***

is a digital measure of the light exposure of each pixel, corresponds to the incoming light intensity. The gray value is linearly related to the mean intensity of the scattered laser light received by the pixel. The images of the particles appear on the digital recordings as isolated bright spots (higher light intensity) on a dark background as shown by Fig. 5.3. The digital images, also referred to as *frames*, have a horizontal and vertical resolution that is equivalent to the number of pixels that make up the photo-sensor.

The PIV recording is the mapping of light intensity from the object plane containing the illuminated seeding particles onto the image plane (CCD photo-sensor) which is then converted into a digital image field (Fig. 5.4). The position of the particles in the frame corresponds to local high light intensity which has a high pixel gray value relative to the background (ie, low pixel gray value). Localized high pixel gray values are referred to as *particle images* where the *particle image diameter* (d_τ) measured in pixel units is related to the number of pixels that are illuminated across the isolated bright spots. Digital recognition of the location of the particle images on the image plane allows for automated computation of the average displacements of *local groups* of particle images to be recorded on the pair of frames.

5.2.1 PIV Methods

There are two primary methods to implement PIV:

1. Double exposure/single frame PIV, where the particle displacement vectors are based on statistical analysis algorithms that employ spatial auto-correlation;

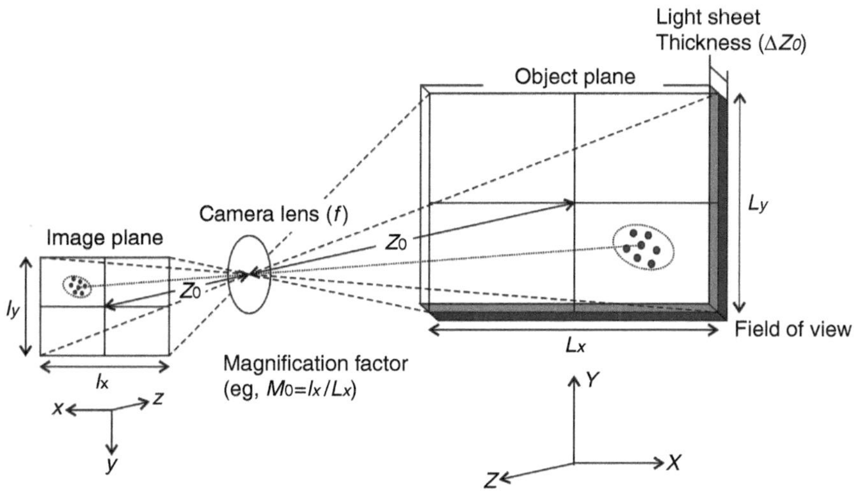

Figure 5.4 ***Schematic illustrating how the object plane is mapped into and image plane coordinates.***

2. Single exposure/double frame PIV, illustrated in Fig. 5.5: numerically calculating the particle displacements using spatial cross-correlation.

The latter is preferable because auto-correlation has an inherent characteristic where the displacement vectors have 180° directional ambiguity, which necessitates additional techniques to determine the true displacement vector direction (sign). Implementation of spatial cross-correlation methods removes this limitation as a result of recording the location of seeding particles in the measurement area on successive frames (Willert and Gharib, 1991; Keane and Adrian, 1992; Westerweel, 1997; Adrian and Westerweel, 2011).

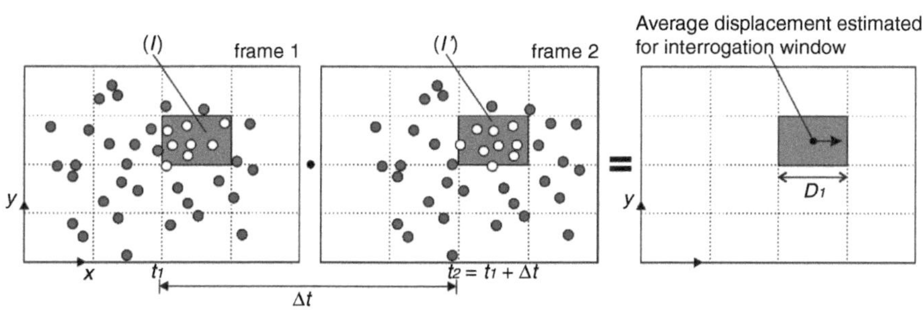

Figure 5.5 ***Schematic illustrating how the image frames are subdivided into interrogation windows (I, I') and a single displacement vector is estimated for each interrogation window.***

The single exposure/double frame PIV method involves capturing two images of the randomly dispersed seeding particles. These images are separated by the time interval (Δt) that is equal to the time between the light sheet pulses. Consequently, this requires that the laser is capable of firing two pulses of laser light in the visible light spectrum and in close succession to one another. This is typically achieved with PIV-specific dual-cavity, frequency-doubled lasers (Bryanston-Cross et al., 1992; Raffel et al., 1998). These types of lasers have two isolated laser cavities that are contained within the laser unit and can be fired independently. The laser cavities initially emit *invisible* infrared radiation (eg, 1064 nm wavelength) that passes through a frequency doubling crystal and is converted to visible green laser light (eg, 532 nm wavelength) (Adrian and Westerweel, 2011). The path of the two laser beams is combined inside the laser unit and exits the laser unit through a single aperture.

Similarly, PIV has special requirements of the CCD camera that allows a pair of images to be acquired in close succession. CCD camera technology has been developed for PIV applications to enable dual image acquisition so that automated cross-correlation of the separate images can be carried out (Raffel et al., 1998; Wernet, 2000; Adrian, 2005). With modern PIV CCD cameras, a time interval of the order of 200 ns between frames is achievable.

5.2.2 Interrogation

The determination of the average displacement vectors of local groups of particles is obtained when the PIV recordings are *interrogated* using the data acquisition and processing software of the PIV system. The interrogation procedure is the implementation of the spatial cross-correlation algorithms that are built into commercially available PIV systems. Herein, the working principle of the PIV spatial cross-correlation method is provided, in brief, to aid in the setup of the PIV system to produce PIV recordings that are amenable to interrogation.

Initially, the pair of frames is divided into subregions, referred to as *interrogation windows*, of a certain size (D_I) that is measured in units of pixels (eg, 32 × 32 pixels). Within a given interrogation window of the *first* frame (I), a certain number of particle images are captured. In the subsequent *second* frame, the same group of particle images should appear in the corresponding interrogation window (I'), but the position of the group of particles with respect to the first frame will have changed if the fluid carrying the seeding particles is moving, as shown in Fig. 5.5.

The size of the interrogation windows (D_I) as well as the time interval (Δt) between frames are chosen so that the loss of pairs of particles between the two corresponding interrogation windows due to natural particle motion is minimized. Loss of pairs is where particles enter or leave the subregion through either (1) in-plane motion because the size of the interrogation window is too small or the time interval too large or (2) out-of-plane motion that is caused by strong 3D flow behavior.

Importantly, PIV interrogation is based on the assumption that the size of the interrogation window is small enough so that the local velocity gradients inside the subregion are insignificant. Therefore, the particle images appearing within the interrogation window have uniform velocity as there is negligible relative motion between the particles. Hence, an average displacement is calculated for each interrogation window in the frame. However, this assumption leads to inaccuracies in regions of the flow field that experience high levels of shear (velocity gradients) such as within the boundary layer and, therefore more advanced methods of interrogation are required (Westerweel, 2008; Adrian and Westerweel, 2011; Wilson and Smith, 2013). The average velocity is evaluated by dividing the local displacement of each subregion by the time interval (Δt).

Spatial cross-correlation measures the degree of "match-up" between two interrogation windows when one subregion is spatially shifted relative to the other. Essentially, an interrogation window is a 2D map of discrete image intensities (pixel gray values), representing particle image locations (ie, match-up of particle image locations). Mathematically, a 2D discrete cross-correlation function provides the quantitative measure of how well the two subregions overlay each other when one region is shifted by integer amounts with respect to the other.

The cross-correlation values that are a function of the discrete spatial shift variables (x, y) may be expressed as:

$$\phi_{II'}(x,y) = \sum_{i=-\infty}^{\infty} \sum_{j=-\infty}^{\infty} [I(i,j)][I'(i+x,j+y)] \tag{5.1}$$

where i and j give the positions of the interrogation windows where subsampling occurs. The function (Eq. 5.1) is evaluated by the summation of the products of the pixel gray values in the two interrogation windows. A peak cross-correlation value results when the interrogation windows of the first and second frames match-up. The numerical value of the spatial shift variables that results in the best match-up (ie, the highest correlation values) is a direct estimate of the average displacement vector for the group of particles that were captured in the interrogation windows. Hence, detecting the location of the so-called *displacement correlation peak* in the cross-correlation plane determines the average displacement of the particles in a subregion as shown in Fig. 5.6.

The location of the peak or maximum value of $\phi_{II'}(x, y)$ in the correlation plane is equal to the average displacement of the group of seeding particles recorded within a given interrogation window. Fig. 5.6 indicates that the displacement correlation peak is shifted by, for example, approximately eight pixel units along the x-axis. Therefore, the average displacement of the particles in the image plane is also eight pixels along the x-axis.

In practice, interrogation procedures used in commercial PIV systems typically perform the cross-correlation calculations in the frequency domain utilizing Fast Fourier Transforms (FFTs). FFTs simplify and reduce the computational expense of interrogation compared to the discrete cross-correlation function (Eq. (5.1)). However, the location of the displacement correlation peak is still resolved in terms of the spatial shift variables (x, y) as an inverse FFT is carried out after the cross-correlation calculation. Furthermore, the cross-correlation values resulting from Eq. 5.1 are typically normalized, which allows comparisons about the degree of match-up between different interrogation windows to be made, where the peak cross-correlation coefficient value approaches unity (Willert and Gharib, 1991; Raffel et al., 1998).

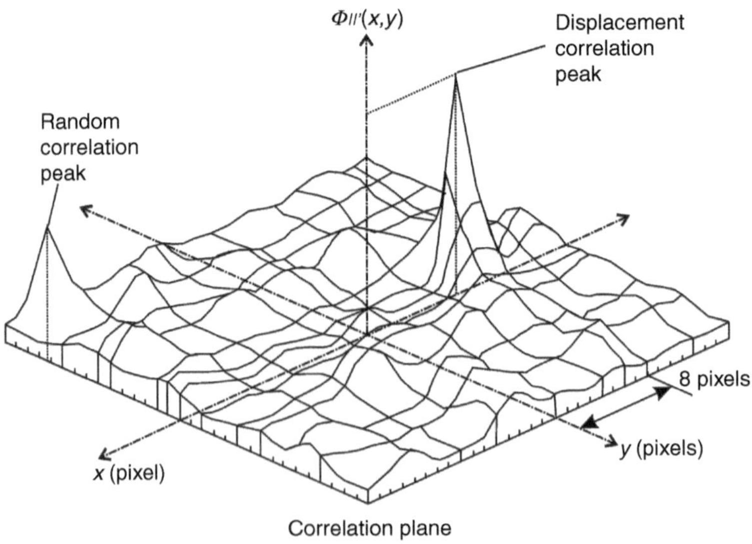

Figure 5.6 ***Schematic of the displacement correlation peak in the correlation plane.***

The amplitude and width (in pixel units) of the displacement correlation peak are strongly influenced by the setup of the PIV subsystems. These peaks need to be clearly detectable over and above the peaks related to random correlations caused by noise recorded within the digital images as shown in Fig. 5.6. Crucially, the reliability of velocity field calculations is related to the *valid detection* of the displacement correlation peak. Hence, the setup of the PIV system must be tuned to maximize the amplitude of the displacement correlation peak. A peak detection criterion typically used is that the amplitude of the displacement correlation peak should be at least 1.2 times greater than the amplitude of the random correlation peaks (Adrian, 1991; Keane and Adrian, 1992). An image mismatch or the loss of particle pairs between corresponding interrogation windows reduces the probability of valid peak detection by the PIV interrogation algorithm. In such a case, random correlation peaks may be falsely identified as the displacement correlation peak, causing erroneous or spurious vectors to be calculated for a given interrogation window.

Peak detection and measurement reliability are also compromised if the number of particle images in the interrogation window (N_I), referred to as *image density*, is insufficient. The amplitude of the displacement correlation peak is directly proportional to the image density provided that a fixed number of particle pairs remain within the subregion. The image density is controlled directly by the concentration of the seeding particles introduced into the fluid flow.

A reduction of the displacement peak amplitude and an increase of peak width are also caused by velocity gradients within the interrogation window. The effects of velocity gradients are significant when the local displacement variation within the interrogation window is greater than the mean particle image diameter (d_τ). Substantially larger displacement variations eventually cause the displacement correlation peak to split into numerous correlation peaks. Thus, local velocity gradients also adversely affect the valid detection of the displacement correlation peak and introduce a certain degree of bias error into the velocity field measurement (Westerweel, 2008).

To improve the accuracy of the displacement estimates, the profile of the displacement correlation peak is approximated using an appropriate curve fit, such as a three-point Gaussian fit (Westerweel, 2000). The approximation, therefore allows for a more precise estimate (within fractions of a pixel) of the location of the displacement correlation peak on the correlation plane using *subpixel interpolation*. Consequently, the width of the displacement correlation peak is an important parameter to ensure that the fitted curves closely approximate the shape of the displacement correlation peak because a poor curve fit contributes to displacement measurement error (Westerweel, 2000).

It has been established that the width of the displacement correlation peak is proportional to the mean particle image diameter, provided the particles within the interrogation windows have uniform displacement (velocity) (Wernet and Pline, 1993; Westerweel, 2008). In addition, theoretical and numerical studies have demonstrated that the displacement measurement error is also directly proportional to particle image diameter (d_τ) when (d_τ/d_r) > 2 and the image density (N_I) is fixed where d_r is pixel size (Westerweel, 2000). The particle image diameter (d_τ) can be directly controlled in a number of different ways, for example, (1) seeding particle diameter (d_p), (2) the magnification factor (M_0) between the object plane and the image plane, and (3) the aperture setting or *f*-number ($f^{\#}$) of the CCD camera. Therefore, a strong link is established between the accuracy and reliability of the velocity field estimate and the setup of the PIV subsystems.

5.3 PIV SYSTEM PARAMETERS

The accuracy and reliability of the velocity field calculation are strongly dependent on the quality of the PIV recordings that require maximum visibility of the particle images. Visible particle images are related to the mean light exposure of the pixel, which is dependent on a number of parameters relating to the various components of the PIV system. In general, the most easily adjustable parameters that can alter the mean pixel exposure include:

1. The laser pulse energy (J_0),
2. The camera aperture size (D_a) or *f*-number ($f^{\#}$),
3. Light sheet thickness (Δz),
4. Particle diameter (d_p), and
5. CCD camera magnification factor (M_0).

Unfortunately, one parameter does not operate in isolation of the others and altering one may require the other parameters to be adjusted. Hence, setting up the PIV system to acquire high-quality images for interrogation is an iterative process that requires refinement. A few recommendations aimed at improving the quality of the images to aid interrogation are provided, classified according to subsystem.

5.3.1 Flow Test Rig

The flow test rig provides the flow boundaries to a typical fluid flow experiment. As PIV is an optical measurement technique, optical access to the fluid flow through the flow boundaries is required. The walls of the test rig need to be constructed from a transparent material (eg, glass or Perspex/Plexiglass). In planar or 2D PIV experiments, a single CCD camera is positioned perpendicular to the plane of the laser light sheet. Consequently, the transparent walls should be placed at right angles to each other: (1) to provide access into the test section from an external laser light sheet and (2) to allow the scattered light from the seeding particles to be received by the CCD camera that is positioned outside of the test rig. The transparent walls should be thin to minimize optical distortion of the light that is transmitted through the flow boundary.

The other background surfaces of the test rig should be masked (eg, using matte black covering) to improve the visibility of the seeding particles on the PIV recordings. Masking also limits the ambient light entering into the test section so that only the laser illumination is transmitted to the CCD photo-sensor. Taking the PIV measurements in a dark room also improves the quality of the images.

Laser illumination reflected from highly reflective surfaces can cause saturation or blooming of the CCD camera which can also permanently damage the photo-sensor. Areas that cause surface flaring on the PIV images are also unable to be interrogated. Surfaces with a high level of reflectivity should be masked with a black paint or adhesive or even constructed of a transparent material (eg, glass or Perspex) which has very low reflectivity.

5.3.2 Seeding Generator

In wind tunnel-based PIV experiments, a seeding generator is commonly used to introduce the seeding or tracer particles into an airstream. The type of seeding generator used is dependent on the seeding material that is considered to be most appropriate for the PIV experiment and test facility.

Fluid-based seeding generators create fine liquid droplets by atomization. Special atomizer chambers filled with liquids (eg, olive oil, vegetable oil, or a mineral oil) employ either the open jet pump principle or the Laskin nozzle to create small particles of the order of 1–2 μm in diameter (d_p). A mist of seeding particles can also be created by vaporizing the seeding fluid with a local heat source. The air flows can also be seeded with Expancel, hollow polymer spheres that have a particle size of 5–100 μm and density of 20–40 kg/m^3. The polymer spheres require stirring and aeration when they are dispensed into the airstream to ensure that the particles are uniformly dispersed. The ability of Expancel to closely follow the fluid flow is reported by Adrian and Westerweel (2011) to be less than desirable due to the relatively large particle sizes compared to liquid droplets formed by atomization, so it may only be appropriate in low-velocity fluid flows. Atomized liquid droplets or other types of seeding particles also settle on surfaces of the test facility and equipment, leaving residue that may require considerable effort to clean off. Covering optical and other specialist equipment (eg, mirrors or camera lenses) when not being used for data acquisition may reduce their risk of damage.

5.3.2.1 Seeding Particles

The seeding particles have two functions that are conflicting in nature. First of all, seeding particles act as markers to the fluid flows that are required to closely follow the flow pattern. Small particles with a density that matches the airstream (neutrally buoyant) are most desirable because they have the best frequency response characteristics to the fluid forces that act on the particles. The error introduced from particle slip where the particle lags behind the flow is minimized with small, low-mass particles. In practice, it is difficult to seed air flows with particles that have the same density and particles tend to be denser than the surrounding fluid. The implication is that the particles will underrespond to flow field accelerations, although with incompressible subsonic flow, this lag error is considered small (Wernet, 2000). The second function requires that the seeding particles are clearly visible on the PIV recordings. Larger particles are more effective at scattering the laser illumination to provide the greatest level of exposure at the image plane. Increased exposure from the seeding particles enhances the detectability of the displacement correlation peak.

The light scattering performance and hence the pixel exposure are also strongly dependent on the refractive index (m) of the seeding material relative to the fluid. Increasing the refractive index by selecting a suitable seeding material is an alternative way to increase the pixel exposure level and compensate for smaller particle sizes.

A certain seeding concentration is needed for the cross-correlation to work. Keane and Adrian (1992) have demonstrated that to ensure 95% probability of valid peak detection the image density, the number of particle images in the interrogation windows, should be $N_I > 6$. Westerweel (2000) has also confirmed with theoretical and numerical studies that the measurement error becomes independent of image density when $N_I > 5$. Measurement reliability is dependent on ensuring that a sufficient number of particle pairs are in the interrogation windows. Adjustments to the quantity of seeders produced by the seeding generator may be done iteratively, gradually increasing the seeding concentration until the target value of seeders is established and interrogation results are satisfactory.

Generating a high enough concentration of seeders for correlation is often problematic when conducting wind tunnel-based PIV experiments. Recirculating the air flow that is carrying the seeding particles using a closed-circuit wind tunnel has the benefit of increasing the seeding concentration gradually without the need for a large-volume seeding generator.

In general terms, the seeding concentration needs to be uniformly dispersed across the measurement area. An ideal concentration results in a high image density so that it is not possible to visually distinguish fluid flow features from the PIV images. The flow features are revealed after the interrogation process when the velocity field is calculated. Hence, PIV seeding concentrations are inherently homogeneous. The situation is different when seeding for flow visualization experiments that deliberately use inhomogeneous seeding concentrations to make flow structures visible. In these, seeders are introduced into a local region of the flow field where flow structures need to be observed.

The correlation calculations are also most reliable when the seeding is mono-dispersed, where the particles have approximately the same diameter and shape. Poly-dispersed seeding hinders cross-correlation, because wide variation of particles sizes adversely affects the interrogation of the images. In such instances, additional filtration to remove particles of a certain size is beneficial. This can be achieved using a cyclone separator or centrifuge that is effective at removing excessively large seeding particles from the seeding source.

5.3.2.2 Seeding Material Health Hazards

Consideration must be also given to the health hazards of using certain materials as flow seeding agents because some seeding materials may be toxic or carcinogenic when inhaled. The appropriate personal protective equipment should always be worn, for example, breathing apparatus. Adequate room ventilation systems should be installed into the test facility.

5.3.3 Pulsed Laser System and Light Sheet Optics

The laser system is necessary to illuminate the seeding particles. The most crucial parameters that affect the image quality and, therefore the reliability of the velocity field estimates are:

1. Laser energy,
2. Overlap of the two light sheets,
3. Time interval between pulses, and
4. Light sheet thickness.

In the interrogation analysis, it is of vital importance that the particles that were captured in the first frame are again recorded in the next. Furthermore, a specific particle can only be recorded if illuminated. It is in this context that the setup of the laser illumination system is examined.

The failure of a seeding particle to be illuminated can be a result of two factors:

1. There is simply not enough laser energy in the light sheet to scatter sufficient light intensity towards the photo-sensor of the CCD camera.
2. The two light sheets emitted from the dual cavity are not overlapping.

Lack of particle illumination and pixel exposure can be compensated for by increasing the energy of the laser light sheet emitted from the laser unit. This will increase the level of exposure on the CCD photo-sensor. In the situation where the laser light sheets from the two laser cavities are not overlapping, careful adjustments must be carried out on the laser unit to align one of the laser light sheets with the other. The light sheet alignment is necessary to ensure that the same plane is illuminated when the laser pulses so that the same group of particles is made visible in successive frames. Illuminating different groups of particles will lead to invalid detection of the displacement correlation peak; consequently, random or spurious velocity field estimates could result.

5.3.3.1 Laser Light Sheet Alignment

A detailed procedure on how to adjust the laser light sheet alignment is usually included with the laser's operators manual. Here, we suggest methods to accurately assess the alignment of the two laser light sheets. The most straightforward method is to aim the laser at a target that is made from photosensitive paper (eg, ZAP-IT laser alignment paper) that will change color or burn when exposed to the high-intensity light source. Iterative adjustments can be made to the beam combining optics until the two light sheets overlap at a set distance between the laser and the target. Alternatively, the laser unit is aimed at a black target surface where the distance between the laser and the target is comparable to the PIV experimental setup. The CCD camera (synchronized to the laser pulses) is used to take images of the two light sheets. Permanent damage to the photo-sensor can occur if the CCD is overexposed to the high-intensity laser illumination. To prevent possible damage, strong neutral density filters should be placed in front of the camera lens to attenuate the laser light intensity being received by the CCD. When digital images of the two light sheets are recorded, the intensity profiles (pixel gray values) across the light sheet thickness can be accurately overlaid by making iterative adjustments.

5.3.3.2 Laser Pulse Timing

The time interval (Δt) between successive laser pulses is an important parameter because it can set the mean particle displacement within an interrogation window. The objective when setting the time interval between successive pulses is to minimize the loss of particle pairs in the interrogation window through in-plane motion. Westerweel (1997) advocates design rules that provide guidelines to ensure an optimal setup of the time interval between laser pulses, which is also coupled to the actual size of the interrogation window (D_I) (unit: pixels). The time interval should be set so that the mean in-plane particle displacement (unit: pixels) in the interrogation window is no greater than a quarter of the interrogation window size (ie, $\Delta x \leq D_I/4$). When estimating the mean in-plane displacement, a certain degree of prior knowledge of the velocity field is necessary to gain an initial estimate of the maximum particle velocity on the object plane. Based on the initial estimate, the time interval can be calculated given the maximum displacement of seeders in the interrogation window allowed by the ($\Delta x \leq D_I/4$) design rule. Mapping particle displacements from the object plane to the image plane is related to the magnification factor (M_0) that is determined when setting up the CCD camera. The time interval should also be adjusted iteratively until the outcome of the interrogation process is satisfactory.

5.3.3.3 Laser Light Sheet Thickness

The loss of particle pairs between corresponding interrogation windows due to out-of-plane motion can be limited to a certain extent by increasing the laser light sheet thickness (Δz). Valid displacement correlation peak detection relies on having particles that were captured in the first frame remain in the second frame. Another design rule recommends that the out-of-plane motion of the seeding particles should be less than a quarter of the light sheet thickness. However, increasing the light sheet thickness has other implications that may not be favorable to producing high-quality images for interrogation, such as reduced pixel exposure requiring higher laser energy to compensate. To accommodate the thicker light sheet, the depth of field should also be increased, by reducing the CCD camera lens aperture size (increasing the $f^{\#}$). The depth of field is the distance (ΔZ) either side of the object plane where particles remain well focused. The sheet thickness is usually adjusted on the light sheet forming optics attached to the laser source.

5.3.3.4 Laser Safety

Those who are unfamiliar with the significant health risks of operating a high-energy pulsed laser should consult the appropriate laser safety guidelines about using class IV lasers before any PIV experiment is executed. The setup of the experiment and test facility should incorporate all laser safety precautions (such as warning signage, access control, laser system interlocks, and laser curtains) to ensure the safety of the laser operator, colleagues, and visitors. Always wear the appropriate laser eye protection that is suitable for the laser used in the PIV experiment, for example, laser safety goggles for a 532 nm wavelength laser.

5.3.4 CCD Camera System

The CCD camera is the device that transforms incident scattered light from the seeding particles into digital particle images. For planar PIV, it is essential that the camera's optical axis is perpendicular to the light sheet. The two most important setup parameters concerning the CCD camera are (1) image magnification factor and (2) particle image diameter.

The image magnification factor (M_0) maps the projection of length scales from the object plane that is coincident with the laser light sheet onto the image plane which is the CCD photo-sensor. The scaling relationship is determined during a calibration procedure that must occur before PIV images can be recorded.

5.3.4.1 Calibration

Calibration requires that a calibration plate or target with precisely known dimensions is carefully positioned so that the face of the plate is coincident with the central plane of the laser light sheet. The camera is positioned some distance (Z_0) away from the calibration plate, until the camera's field of view fits the desired measurement area of the PIV experiment. The field of view defines a maximum value of the in-plane location of the particles (Adrian, 1991). Fine adjustments are then made to the camera's focus to bring the markings on the calibration plate into clear view, thus the object plane is established. To calculate the magnification factor, the dimensions of the field of view (object plane) and the photo-sensor (image plane) need to be determined.

The dimensions of the camera's field of view (L_x, L_y) are measured from the images taken of the calibration plate, which have a size typically of the order of 100 × 100 mm^2. The dimensions of the image plane are related to the *pixel format* of the CCD, which is the number of pixels that make up the photo-sensor array (eg, 2048 × 2048 pixels) and *pixel size* (eg, d_r = 7.4 μm) or the pixel pitch between adjacent elements. Hence, the size (l_x, l_y) of the image plane is approximately 15 × 15 mm^2 where, for example,

$l_x = 2048 \times 7.4\ \mu m = 0.01515$ m. The magnification factor (M_0) is the ratio of the corresponding lengths from the object plane to the image plane (ie, $M_0 = l_x/L_x = 15/100 = 0.15$).

Implicit in the use of planar PIV is the "paraxial photography" approximation where images recorded of seeding particles at the object plane fall predominantly within a 2D plane (x, y) and there is a negligible contribution of the out-of-plane displacement of particles (ΔZ) (Adrian, 1991). The approximation is valid when the dimensions of the camera's field of view are much smaller than the distance between the object plane and the camera lens (Z_0) (ie, $L_x/Z_0 << 1$), which implies that the degree of magnification is small. In instances where the flow field has a high level of 3D behavior where out-of-plane particle displacements are substantial, consideration should then be given to Stereoscopic PIV methods. Here, two CCD cameras are aimed at the light sheet. However, their respective optical axes are no longer perpendicular to the plane of the laser light sheet as is the case with planar or 2D PIV methods; more details on Stereoscopic PIV methods can be found in Adrian and Westerweel (2011).

5.3.4.2 Particle Image Diameter

The particle image diameter (d_τ) is an important parameter affecting the accuracy of the displacement estimates. Previous studies have demonstrated that the minimum measurement error has been reported to occur when $d_\tau/d_r \approx 2$ (Prasad et al., 1992; Westerweel, 1997, 2000; Raffel et al., 1998). The particle image diameter can be adjusted using the *f*-number ($f^{\#}$) ring on the camera lens, which is the ratio of the lens focal length (f) to aperture size ($f^{\#} = f/D_a$). In general, increasing the *f*-number ($f^{\#}$) results in larger particle image diameter. However, the smaller aperture size decreases the exposure of the pixel and consequently the particles are not as bright on the PIV images when the $f^{\#}$ is increased. To compensate for this, an increase of the laser pulse energy would be required.

The mean particle image diameter can be determined from the raw PIV images by carrying out auto-correlation of a single frame to determine the correlation peak width (d_c). A symmetrical correlation peak is related to the particle image diameter (ie, $d_c = d_\tau(2)^{1/2}$).

5.3.5 Data Acquisition and Processing

Interrogation requires the choice of a certain size of interrogation window (D_I) (eg, 32 × 32 pixels), that is large enough to minimize the loss of in-plane particle pairs. However, this choice is not made in isolation and is dependent on the time interval between the successive frames and the seeding concentration or image density (N_I). In both instances, the design rules provide a rule of thumb (ie, $\Delta x \leq D_I/4$ and $N_I > 6$) to ensure the cross-correlation calculations yield accurate and reliable estimates. So, in effect, the choice of interrogation window size is set with the time interval and the seeding concentration, all of which are embedded within the PIV recording. Making the choice to reduce the size of the interrogation window (eg, from 32×32 pixel^2 to 16×16 pixel^2) will require a new set of images with a smaller time interval and a higher seeding concentration to maintain the required number of particle pairs in the interrogation window for cross-correlation calculations.

The benefit of reducing the interrogation window size is that the spatial resolution of the velocity field increases. Increasing the spatial resolution results in more velocity vectors within a single frame. This reveals a greater level of detail about certain flow structures. However, there is an evident compromise between spatial resolution (smaller interrogation windows) and measurement reliability that calls for larger interrogation windows to limit the loss of in-plane particle pairs, as indicated by the $\Delta x \leq D_I/4$ design rule. An advanced interrogation approach, described by Westerweel et al. (1997), has been developed to overcome this apparent compromise by using interrogation window offsetting.

By using interrogation window offsetting, both a higher spatial resolution and a higher maximum displacement of seeder particles within an interrogation window can be achieved. Peak detectability for the displacement correlation peak is optimized as a result of reducing the loss of in-plane particle pairs. The procedure for using interrogation window offset is to first conduct an initial PIV interrogation analysis to obtain an estimate of the velocity field within the measurement area. Based on the local mean displacement vectors of each interrogation window, the corresponding interrogation window is offset or shifted by the nearest pixel unit, so the particle images effectively overlap and it is the fractional pixel displacements that are then determined in the following interrogation procedure. The instantaneous velocity field is estimated when two successive frames are interrogated over the time interval (Δt). Time-averaged or average correlation velocity field information also can be obtained by recording an ensemble of instantaneous measurements over a long period of time. A sequence of single exposure/double frame images needs to be recorded and then the time-averaged correlation is estimated for a given interrogation window and for a large number of frames. The sampling rate for the sequence of single exposure/double frame images is typically of the order of 5~10 Hz.

Image processing can improve the quality of the PIV recording before interrogation, by removing random noise recorded in the images. There are many ways to process the images to make them more amenable to interrogation and to reduce the likeliness of random or spurious vectors emerging from the calculations. One method is *image background removal* that aims to remove noise caused by illumination intensity fluctuations which may be due to natural instability of laser sources or reflections or illumination obstructions from large seeding particles in the optical path of the light sheet. One way to implement image background removal is to use pixel-by-pixel normalizations (Adrian and Westerweel, 2011; Koschatzky et al., 2011). Here, the time-averaged pixel gray value for a given pixel in the frame over a number of recordings is determined first. Effectively, a "mean" image is generated where each pixel has a time-averaged gray value. Each image in the ensemble is then normalized by the mean image. This has the benefit of removing noise within the images that generates random correlation peaks. This practice improves the number of valid peak detections of each frame pair.

Image masking is also an important way to remove regions of the image that are underexposed due to low light sheet intensity or subjected to surface flaring from laser light reflections. Such regions adversely affect the cross-correlation calculation of the particle images. Therefore, it is beneficial to manually mask parts of the image that are likely to interfere with the success of the interrogation procedure.

Example 5.1: Horseshoe Vortices Around an Inclined Cylinder

A. Background

The boundary layer approaching an obstacle protruding from an endwall experiences an adverse pressure gradient in the direction of the mainstream. Subsequently, the boundary layer on the endwall forms a vortical structure, the so-called "horseshoe vortex" around the junction of the obstacle and the endwall. In accordance with Kelvin's circulation theorem, the boundary layer vorticity cannot be destroyed. Instead, it is convected around each side of the obstacle to form the two legs of the horseshoe vortex. A considerable amount of work on different aspects of the horseshoe vortex has been conducted (Seal et al., 1997; Visbal, 1991; Hung et al., 1991; Baker, 1980; Ballio et al., 1998; Simpson, 2001; Khan and Ahmed, 2005; Khan et al., 1995; Tobak and Peake, 1979; Hunt et al., 1978; Perry and Chong, 2003). It has been numerically and experimentally revealed that the horseshoe vortex system can consist of multiple pairs of counter-rotating vortices which undergo cyclic-periodic events of their formation at high Reynolds numbers (Seal et al., 1997). The exact number of vortex pairs depends on flow conditions (Visbal, 1991).

Topologically, the axial upstream location of the primary separation point (also called the "saddle point") in the plane of symmetry for a turbulent flow decreases with decreasing displacement thickness of the incoming flow (Hung et al., 1991).

It has been experimentally quantified in the turbulent flow regime (Baker, 1980) that when flow parameters (eg, Reynolds number and cylinder diameter to displacement thickness ratio, D/δ^*) are fixed, the longitudinal location upstream from the cylinder axis is largely constant in the plane of symmetry if the cylinder is relatively long ($l/D \geq 1.0$) with $S_{\text{primary}} \sim 1.1D$ and $S_{\text{secondary}} \sim 0.8D$ in the two-vortex system.

A variety of topological models of the horseshoe vortex system in the plane of symmetry upstream of the obstacle have been proposed (Visbal, 1991; Khan and Ahmed, 2005; Khan et al., 1995; Tobak and Peake, 1979; Hunt et al., 1978; Perry and Chong, 2003). For a wide range of Reynolds numbers, the topological transition from a single-vortex system to a three-vortex system (ie, three pairs of counter-rotating vortices) and its oscillatory behavior between three- and four-vortex systems have been identified. Three distinct modes which are mainly classified by Reynolds numbers show steady, oscillatory, and shedding and splitting modes. Although many have devoted effort to gain fundamental insight into the topological behaviors of horseshoe vortices, existing works have been limited to studying the horseshoe vortex system formed around obstacles placed in the upright position.

This example aims to elucidate effects of cylinder inclination on a turbulent horseshoe vortex system around the cylinder–endwall junction when the circular cylinder is backward inclined (Fig. 5.7). To this end, particle image velocimetry (PIV) measurements were carried out in a wind tunnel. The inclination angle was varied from $\alpha = 0$ to 45° with a 15° increment and the Reynolds number was fixed at $Re_D = 7.8 \times 10^3$ during the entire experiment.

B. Experimental Details

An open-type wind tunnel was used for flow field measurements using PIV. A centrifugal fan blows air that passes through a settling chamber, screens, and a honeycomb section. The size of the rectangular test section is 0.12 m (height) × 0.14 m (width).

Single circular cylinders with inclination angles of $\alpha = 0$, 15, 30, and 45° were fabricated for the testing. The inclination angle α is measured from the z-axis, which is normal to the mainstream flow direction. Positive inclination angle indicates clockwise rotation with respect to the y-axis (along the channel width). Thus, the cylinder is rotated clockwise as its inclination angle is increased.

Typically, two different methods can be used to change the inclination of a cylinder. One is to fix both ends of the cylinder to the endwall plates.

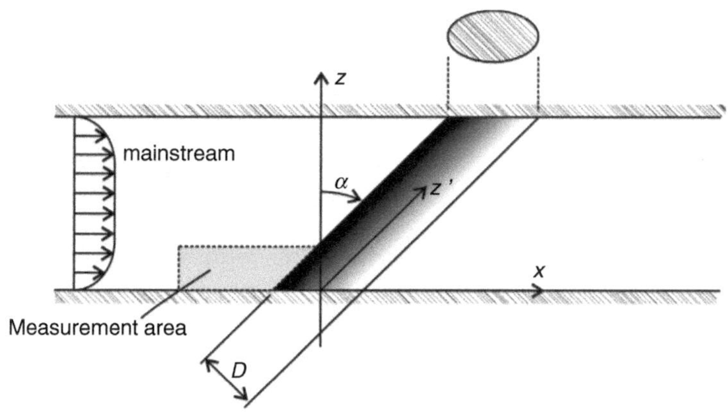

Figure 5.7 ***Configuration of a single cylinder with inclination.***

When the lower endwall plate is shifted longitudinally and upward, the angle between the mainstream and the cylinder axis increases. During this inclination, the cylinder length remains unchanged. Another way, which is employed in the present example, is to fix the channel height (H) and rotate the cylinder with respect to the y-axis. The consequence is that the length (l) of the cylinder sandwiched between the channel endwalls is increased according to $l = H/\cos(\alpha)$.

A PIV system composed of a double pulsed Nd:YAG laser generator, a CCD camera, and a seed particle generator was used to observe the vortical flow patterns around the cylinder especially at the junction between the cylinder and the bottom endwall plate. A 1 mm thick laser sheet was created by an optical system and aligned with the cylinder axis ($y = 0$) for the x–z plane flow measurement (Fig. 5.7). Prior to the measurement, the endwall plate was painted black to minimize reflection. Based on the flow velocity and the focal length of the CCD camera to the measurement plane, the time interval (Δt) of each pair of captured images (2048 × 2048 pixels) was set to be 80 μs. The instantaneous velocity fields were then obtained using a cross-correlation analysis with an interrogation window of 64 × 64 pixels and a 50% overlap providing vector-to-vector spacing.

C. Physical Interpretation of the Horseshoe Vortex

Fig. 5.8 presents a classical three-vortex system (ie, three pairs of counter-rotating vortices) around an upright cylinder. As previous studies (Visbal, 1991; Khan and Ahmed, 2005; Khan et al., 1995) revealed, the horseshoe vortex consists of multiple pairs of vortices and their locations relative to the cylinder vary in a cyclic manner. According to the graphical illustration reported by (Baker, 1979), at $Re_D = 7.8 \times 10^3$ the three-vortex system may appear with $2 < D/\delta^* < 6$, which is consistent with the present observation.

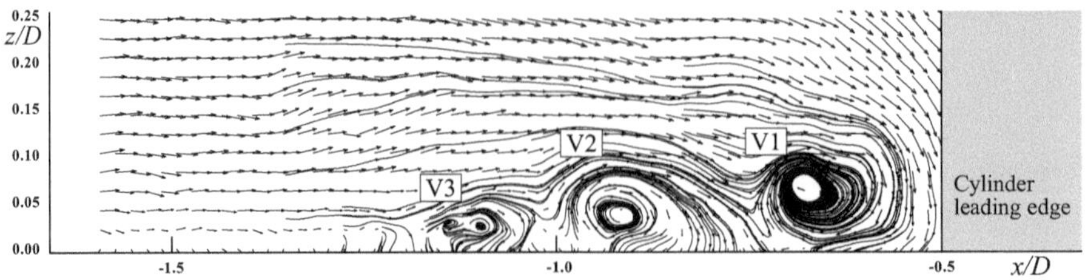

Figure 5.8 ***Instantaneous horseshoe vortex system measured in the plane of symmetry using PIV for an upright single cylinder.***

A vortex (V_1) rotating clockwise is formed immediately upstream of the cylinder leading edge, followed by a second vortex (V_2) and a third vortex (V_3), which are also rotating clockwise. There must exist counter-rotating vortices to satisfy the topology rule (Khan and Ahmed, 2005; Tobak and Peake, 1979; Hunt et al., 1978) although these are not clearly

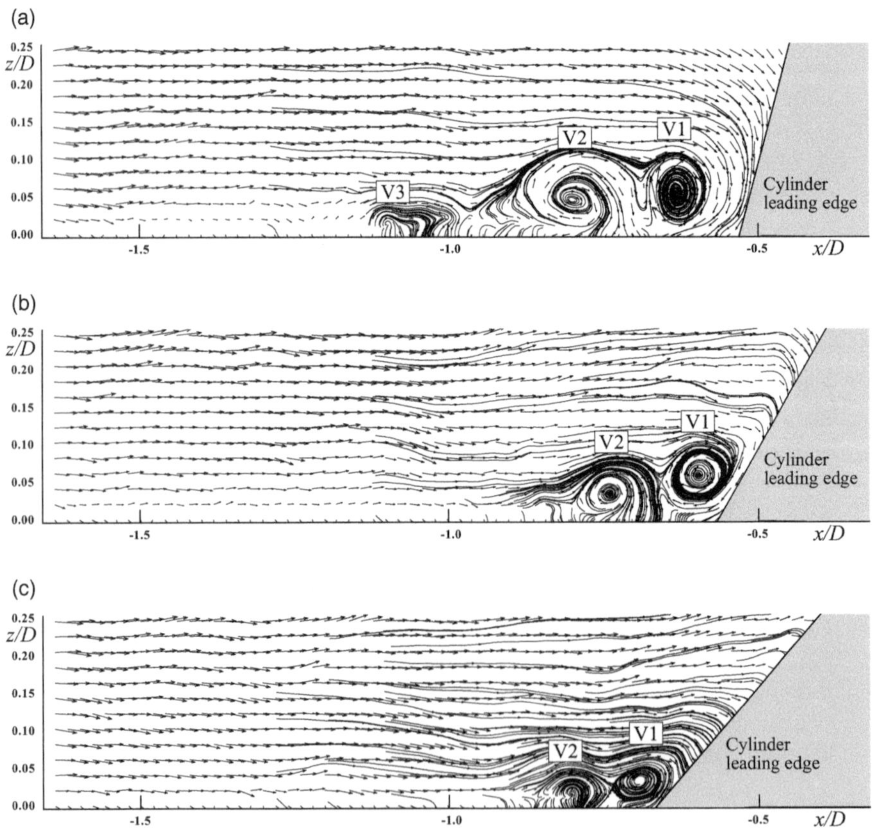

Figure 5.9 ***Instantaneous horseshoe vortex system in the plane of symmetry using PIV for inclined single cylinders.*** (a) $\alpha = 15°$; (b) $\alpha = 30°$; (c) $\alpha = 45°$.

visible in Fig. 5.8. The cores of the vortices V_1, V_2, and V_3 appear to be located approximately at $x/D = -0.7, -0.95$, and -1.2, respectively, from the cylinder axis. Although these locations vary in time and hence are not representative of steady-state features, they at least provide evidence that a three-vortex system is formed in this example. According to a survey of time-averaged characteristics of laminar and turbulent horseshoe vortices (Ballio et al., 1998), the longitudinal position (λ_o) of the primary vortex formed around a cylinder is scattered but located in the range of $0.65D < \lambda_o < 0.8D$, agreeing well with that shown in Fig. 5.8 ($\lambda_o \sim 0.7D$).

In comparison with the upright cylinder case, as the cylinder inclination angle is increased, the vortex (V_1) moves closer to the leading edge of the cylinder as shown in Fig. 5.9. On the other hand, the second vortex (V_2) also moves closer to the leading edge, and further, resulting in a decreased distance between the cores of the two vortices (V_1 and V_2). For larger inclination angles (eg, $\alpha = 30$ and 45°), the third vortex was not recognizable. Such thinning of the horseshoe vortex system is partly associated with the variation of the adverse pressure gradient in the plane of symmetry as the cylinder is increasingly inclined.

Example 5.2: Isotropic-Planar Illumination for PIV Experiments

A. Description of Problem

Velocity field measurements using PIV can only be performed from locations where seeding particles are made visible, typically using laser light illumination. However, PIV also has an intrinsic limitation that stems from how the laser light is introduced into the flow test rig. With conventional PIV methods, the laser light sheet is first formed *externally*. The laser light sheet is then introduced into the flow. If an optically opaque test body is placed in the path of the light sheet, a "shadow" is formed as shown by Fig. 5.10. The seeding particles that are in the shadow region created by the test body cannot be illuminated, which implies that they are not visible on the PIV recordings. Hence, investigators have been limited to regions of the flow field where there are no shadows to obtain PIV velocity field measurements (Sumner et al., 1999; Xu et al., 2003; Wang and Zhou, 2005).

Here, a new method for the illumination of the seeding particles is introduced where the shadow region is eliminated for a particular test setup. Hence, particles that would have previously entered into a shadow are now illuminated and, therefore recorded. This new configuration generates the thin laser light sheet *internally* through the transparent walls embedded with the test body in all radial directions into the flow field. This illumination is isotropic-planar where the laser light is planar and radiates outwards from

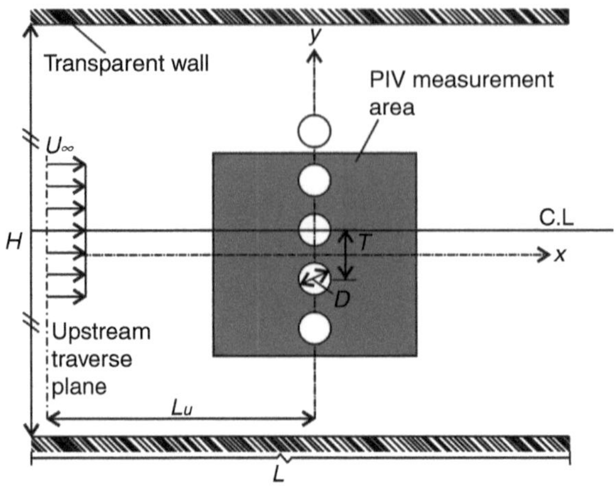

Figure 5.10 ***Schematics of the test section with five side-by-side circular cylinders.***

inside of the test body, creating the so-called "laser disc of light" or "laser light disc." Depending on individual configurations, there is a 360° illumination.

To demonstrate the effectiveness of this new illumination method, PIV measurements of the velocity field around five side-by-side circular cylinders immersed in uniform flow have been conducted (Fig. 5.10). While taking the PIV measurements it was found that the conventional configuration is inadequate. Some regions of the flow field, including the gap region, always have a shadow even if the position of the laser is adjusted in an attempt to eliminate it.

The new method, with 360° illumination around the central cylinder, has allowed the upstream, gap, and downstream flow fields to be measured *simultaneously*. Using this illumination technique we aim to provide physical insight into (1) the development of the upstream flow ahead of the cylinders into the gap flow passing between the cylinders and (2) the interaction of the gap flow with the wake created by the row of cylinders. This configuration has engineering significance that allows for an improved understanding of flow behavior that affects the performance of certain fluidic devices, including heat exchangers where flow mixing is important.

Ishigai et al. (1975) and Moretti and Cheng (1988) have reported that the gap flow behaves like a 2D jet flow that penetrates into the wake region downstream of the row of cylinders. When the spacing ratio is $T/D < 2.0\sim2.3$, instability of the jet has been observed to alter the wake pattern where T is the center-to-center spacing between cylinders and D is the cylinder diameter. The nature of the jet instability is such that the gap flow or jet is deflected towards or away from a particular cylinder in the row. Depending on the direction of the jet deflection a narrow or a wide wake is formed behind the cylinder in question.

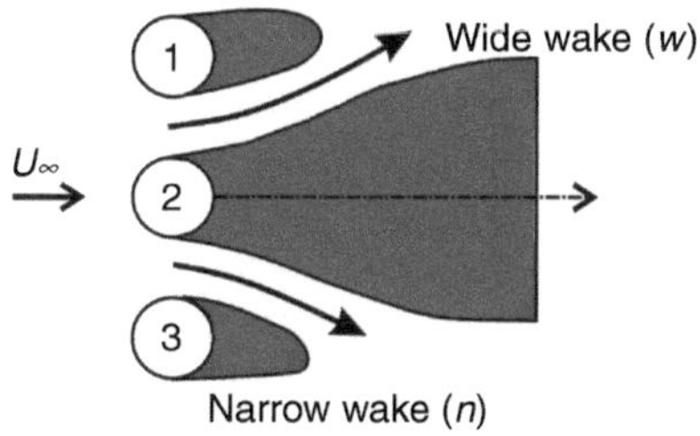

Figure 5.11 ***Schematic of symmetric-biased wake pattern behind three side-by-side circular cylinders*** (Atkins and Kim, 2015).

Sumner et al. (1999) reported on the wake patterns created by three side-by-side circular cylinders (with $T/D = 1.5$). They measured the non-dimensional shedding frequency ($S = fD/U_{\infty}$) immediately behind the row of cylinders where f is the shedding frequency and U_{∞} is the free-stream velocity. They found that the narrow near wake had a higher shedding frequency ($S_n \sim 0.31$–0.33) than the wide near wake ($S_w \sim 0.05$–0.06) that developed behind the central cylinder. The resulting wake pattern is symmetrical about the axis of the central cylinder and the adjacent gap flows are deflected away from the central cylinder, which is classified as the *symmetric-biased* flow pattern (Fig. 5.11).

The PIV measurement conducted by Sumner et al. (1999) showed the wake of a row of three circular cylinders varied with the spacing between the cylinders. However, the flow in the upstream and gap regions was not linked to the observed downstream wake pattern, which may play an important role in affecting the wake pattern. Such a deficiency is primarily due to the shadow region that is created by the cylinders. The PIV measurement setup used by Sumner et al. (1999) may be classified as an external illumination configuration. Simultaneous measurement of the upstream, gap, and wake regions around the row of cylinders may provide further insight into the jet instability that leads to jet deflection.

We compare the velocity fields determined from the conventional external laser light sheet to the new internally generated laser light disc under the same test conditions. In addition, we provide further physical insight into the flow behavior around the row of side-by-side cylinders, specifically, linking the upstream and gap flow behavior to the downstream wake pattern.

B. Experimental Details

B1. Test Rig and Instrumentation

The experiments were conducted in a closed-type wind tunnel with test section dimensions of $W \times H \times L = 0.312$ m $\times$ 0.32 m $\times$ 1.0 m (Fig. 5.10). The walls of the test section are made from transparent acrylic sheet (Perspex) that is 10.0 mm thick. A glass observation window for the CCD camera was inserted into the side of the test section that was 4.0 mm thick. The flow velocity was set to maintain a subcritical Reynolds number based on the single-cylinder diameter of $Re_D = 6000$. Prior to the actual testing, the free-stream turbulence intensity was measured to be $Tu = 1.9\%$ (Dantec MiniCTA 54T42).

Five side-by-side circular cylinders were mounted in the test section (Fig. 5.10). Each cylinder was made from transparent acrylic tube with a diameter of 15.0 mm and a span of 0.312 m. However, for the isotropic-planar illumination setup the central cylinder was made from glass, due to its superior optical properties. The resulting area blockage was 23.5%. The spacing ratio, $T/D = 1.7$, was selected where T is the transverse center-to-center spacing and D is the cylinder diameter = 15.0 mm.

B2. External Illumination

The conventional externally illuminated PIV setup is shown in Fig. 5.12. We note from Fig. 5.12a that the plane of the laser light sheet is parallel to the x–z plane because the base of the laser source is mounted horizontally on a table secured to the floor of the test facility. Consequently, the five side-by-side n

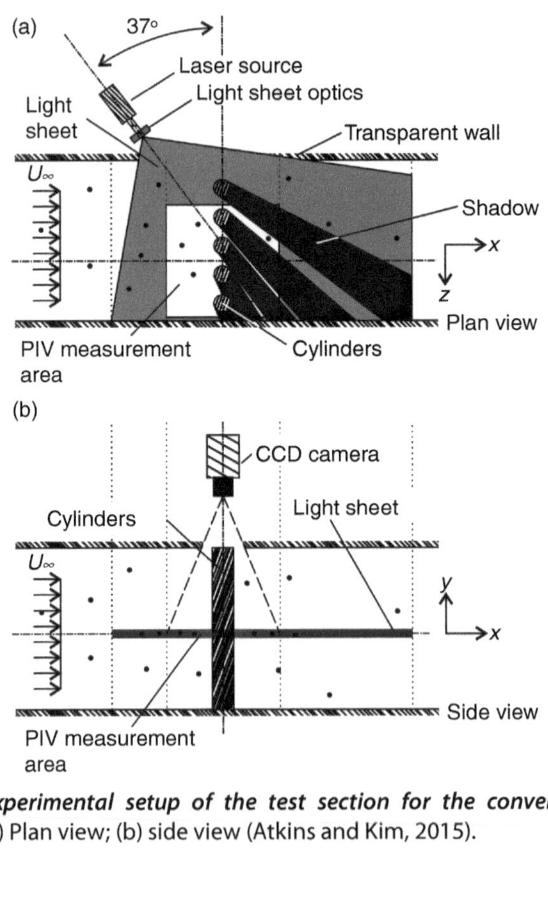

Figure 5.12 ***Experimental setup of the test section for the conventional external illumination.*** (a) Plan view; (b) side view (Atkins and Kim, 2015).

cylinders were mounted so that the cylinder axis is perpendicular to the x–z plane. The optical axis of the CCD camera is also perpendicular to the x–z plane (or parallel to the cylinder axis) for 2D PIV experiments (Fig. 5.12b).

To create the laser light sheet as shown in Fig. 5.12a, light sheet optics is connected to the laser source to transform the emitted laser beam into the laser light sheet. The schematic of the typical light sheet optics shown in Fig. 5.13 illustrates the arrangement of multiple lenses necessary to form a thin laser light sheet. The light sheet thickness (Δy) is manually adjusted (ie, $\Delta y \sim 1.0$ mm) and the laser source energy was approximately 30 mJ/pulse (Dantec Dynamics, 2002).

The objective of the present case study was to quantify the velocity field of the upstream, gap, and downstream regions of the cylinders *simultaneously*. However, it is not achievable using an external light sheet due to the large shadow regions that form (see Fig. 5.12a). The upstream region was, therefore illuminated to observe how the upstream flow field behaves as it approaches the cylinders. The laser is offset by, for example, 37°, which is measured from the transverse z-axis of the five side-by-side cylinders (Fig. 5.12a). This is the maximum angle that could be obtained within the confines of the test facility. Naturally, the angular offset of the laser means that the majority of the gap region and the downstream regions were obstructed by the shadows created by the cylinders as illustrated in Fig. 5.12a.

To circumvent the lack of illumination in these regions, transparent cylinders were also tested to allow the laser light to be transmitted into the shadow regions. However, a significant proportion of the light sheet intensity was attenuated while passing through the transparent cylinders due to refraction and reflection. Consequently, insufficient intensity remained within the light sheet to illuminate the downstream seeding particles.

There are alternative methods that can be used to eliminate the shadow region such as laser beam splitting. With laser beam splitting, the laser

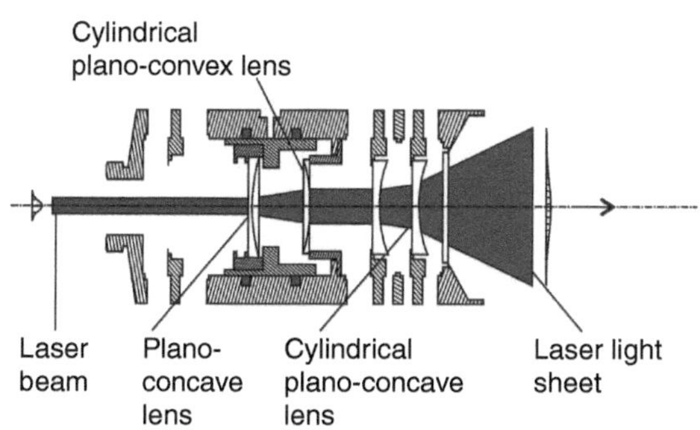

Figure 5.13 ***Light sheet optics, detailing the arrangement of the optical lens required to transform the laser beam into a light sheet*** (Dantec Dynamics, 2002).

beam is split using an optical element and illuminates the row of cylinders from opposite sides. The two separate beams would be directed using planar turning mirrors and light sheet optics placed in the optical path of the beams would transform the laser beam into two opposing light sheets, and hence eliminate the shadow region. The expected limitations of this configuration were primarily due to the complexity of setting up the optical arrangement. A large number of optical elements would be necessary to create the illumination. With an increased number of optical elements, the task of aligning the laser beams and overlapping the opposing light sheets becomes quite difficult and time consuming. There may also be a limit on the available laser source energy as a given amount of laser energy would now need to be distributed over a wider area. Hence, the overall scattered light intensity would be reduced. To compensate, more laser pulse energy would be needed to create the required levels of exposure on the CCD detector and create visible particle images.

B3. Internal Isotropic-Planar Illumination

Internal isotropic-planar illumination is a new optical configuration that was designed and manufactured in-house. The purpose is to eliminate the shadow so that the upstream, gap, and downstream velocity fields of the five side-by-side cylinders can be measured *simultaneously* using PIV.

The shadow is eliminated by introducing the laser light internally through the middle cylinder and radiating the light outwards through the cylinder. To achieve this, (1) the optical axis of the laser beam must be coincident with the axis of the middle cylinder, (2) there must be laser optics inside of the middle cylinder to create the laser light disc, (3) the two beams from the dual-cavity laser must have a circular shape with the same beam diameters and energy, and (4) the middle cylinder needs to be transparent, and ideally would be manufactured from optical glass.

Fig. 5.14 illustrates the experimental setup for the internal isotropic-planar illumination. The optical axis of the laser and the CCD camera is perpendicular to the plane of the laser light disc where the laser light disc is parallel to the x–y plane. The laser source is connected by an "adaptor" to the end of the middle cylinder. As shown from Fig. 5.14b, the middle cylinder is the illuminated cylinder where the planar laser light disc is created and radiates outwards in all directions resulting in 360° laser light coverage around the middle cylinder. Therefore, the upstream, gap, and downstream flow regions are now illuminated simultaneously.

A schematic of the laser light disc optics is provided in Fig. 5.15. A circular, collimated laser beam is emitted from the laser source. This beam is transformed into a laser light disc with a 360° illumination, using a cone mirror with a 90° cone angle.

The fluence of the pulsed laser and the beam profile were characterized and controlled to ensure that the energy density limit or damage threshold

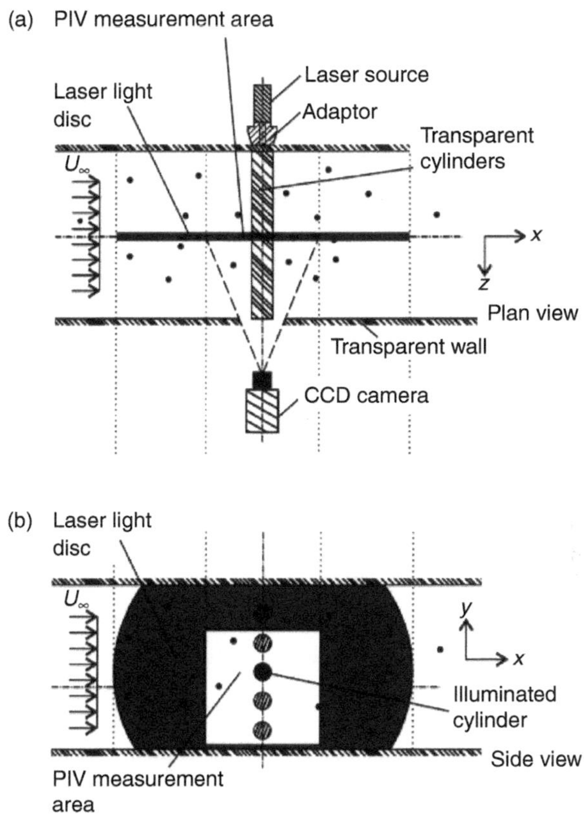

Figure 5.14 ***Experimental setup for the internal isotropic-planar illumination.*** (a) Plan view; (b) side view (Atkins and Kim, 2015).

of the cone mirror (0.2 J/cm^2) was not exceeded. Applying excessive laser energy to increase the illumination of the seeding particles would result in permanent damage to the cone mirror reflective surface. The diameter of the collimated laser beam was approximately 5.5 mm, based on the criterion that the beam diameter is located at the e^{-2} value of the peak intensity. Therefore, the laser source energy was limited to 20 mJ/pulse.

The cone mirror is placed inside an optical shroud. The assembly of the cone mirror and the shroud are then placed inside of the outer clear glass circular cylinder that has a 12.5 mm internal diameter. The optical shroud functions to house the cone mirror and to control the light disc thickness and brightness. The light disc thickness (Δz) is controlled by varying the aperture size.

The brightness is varied by changing the depth that the cone mirror rests inside of the shroud. For a given laser beam diameter and energy level, adjusting the depth of the cone mirror inside of the shroud varies the

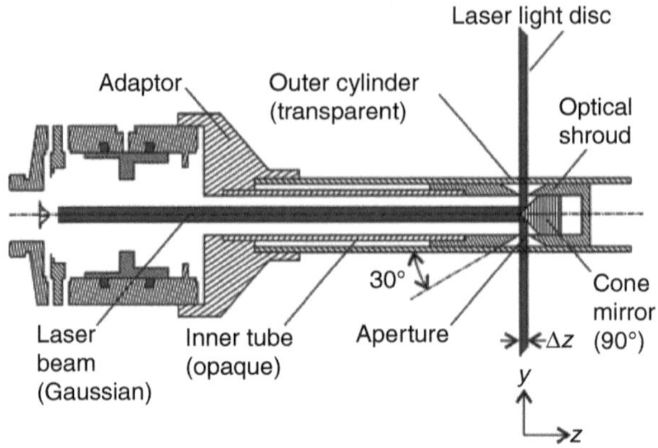

Figure 5.15 ***Laser light disc optics, detailing the arrangement of the adaptor, inner tube, optical shroud, cone mirror, and the transparent outer cylinder which are required to transform the laser beam into a light disc*** (Atkins and Kim, 2015).

energy intensity of the laser light that is reflected from the cone mirror. If it is assumed that the energy intensity distribution of the laser beam has a Gaussian profile (Raffel et al., 1998; Westerweel, 2000), the maximum energy intensity is located at the center of the beam, which then decays exponentially with increasing distance away from the beam center. The brightness of the light disc can be increased if the light that passes through the aperture is reflected from a section of the incident beam that has the highest energy intensity (ie, the beam center). Consequently, the reflected light intensity profile across the light disc is not symmetrical. Referring to Fig. 5.16, the internal isotropic-planar illumination has a normalized intensity profile that is biased towards the side of the laser source. In contrast, the external illumination configuration has a typically symmetrical light intensity profile. However, the biased intensity profile does not appear to affect the recording of the particle images, as high image contrast (ie, visible particle images) is still obtainable.

The aperture size is defined by the gap between the sharp edges of the optical shrouds. The sharp edges have a 30° wedge angle (Fig. 5.15). The wedge angle minimized diffraction of the light disc. A high degree of diffraction is associated with the spreading of the light disc where the thickness increases rapidly as the distance from the central light source increases. The boundaries of the light disc are also blurred, which is undesirable for PIV measurements. The addition of the optical shroud with sharp edges created a desirable light disc for PIV measurements where the boundaries of the light disc are more sharply defined. This can be confirmed from the steep increase of light intensity (normalized gray value) particularly on the laser source side shown in Fig. 5.16. The aperture size should be less than half the incident beam diameter if the masking caused by the sharp edges

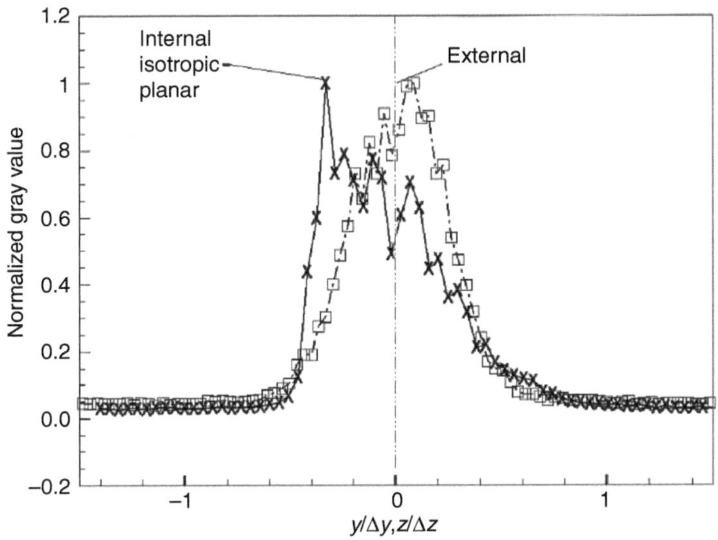

Figure 5.16 ***Normalized laser light intensity profile relating to the external illumination configuration (□) and internal isotropic-planar illumination (×).*** The intensity (pixel gray value) is normalized by the peak gray value (Atkins and Kim, 2015).

is to be effective. In addition, the sharp edges should be positioned so they "cut" the laser light reflected from the cone mirror. The laser light disc thickness was adjusted so that $\Delta z \sim 1.0$ mm.

The laser beam was concealed by an opaque inner tube. The inner tube was necessary to prevent the laser light reflecting from the walls of the outer glass cylinder as the light was transmitted along the cylinder axis. When the inner tube was absent, laser light was reflected and refracted from the transparent walls of the outer tube. The laser effectively illuminated the entire length of the central cylinder, which then caused severe blooming of the CCD camera. Blooming rendered some regions of the PIV recordings unsuitable for interrogation. The end of the inner tube near to the mirror-shroud assembly is supported by a hollow optical shroud with the same internal diameter and wedge angle as the optical shroud that houses the cone mirror. The concealed laser beam also has laser safety benefits, as the laser light is introduced directly into the test section. In contrast, there are significant health and safety risks to the PIV/laser operator that are associated with the external laser light sheet reflecting from the walls of the test rig.

The optical shroud that supports the end of the inner tube also provides a way of supporting additional optical elements, such as a beam reducer or an expander lens. Alternatively, an arrangement of cylindrical lenses may be added to further condition the laser beam before the cone mirror (ie, create a uniform, circular profile).

The experimental configuration (Fig. 5.15) opted to not use additional optical elements (although they have been tested) for the following reasons: (1) the profiles of the pulsed laser beams were largely symmetrical (ie, circular) and (2) the beam divergence angle was small enough to assume that the beam is collimated over the distance that the beam travels to meet the cone mirror.

C. Velocity Field Measurement Using PIV

In both the external and internal illumination configurations, the flow field was seeded with an atomized mineral oil with a mean particle diameter of 1.0 μm. The particles carried by the flow were then illuminated by a frequency-doubled, dual-cavity Nd:YAG laser (New Wave Research™), with a light wavelength of 532 nm (ie, green light). It is assumed that the measured flow field is predominately 2D so that the majority of particles entering into the thin light sheet remain within the light sheet during the measurement period. The measurement period is defined by two sequential pulses of the laser light sheet that are separated by a finite time interval (33 μs).

The recorded flow field is represented by the random pattern of particle images that are mapped onto the image plane of the CCD sensor. The images are recorded on the sensor frame that has a 2048 × 2048 pixel resolution and a pixel pitch of 7.4 μm. The field of view that is set up within the light sheet is shown in Fig. 5.10 where the PIV measurement area is 100 mm × 100 mm. The magnification factor is $M_0 = 0.15$ and the CCD camera lens (60 mm focal length) *f*-number is set (ie, $f^{\#} = 2.8$). The maximum aperture size of the camera lens was used to create the highest pixel exposure and keep the laser pulse energy as low as possible. Due to the damage threshold constraint of the cone mirror, increasing the camera aperture size increases the pixel exposure without risking damage to the mirror. However, a trade-off exists where the depth of field is reduced. To compensate, a sharply defined, thin light sheet is needed so the particle images are well focused and visible on the PIV images.

Image masking was also implemented prior to the velocity field calculations where regions of the flow field that were not suitable for interrogation were excluded from the analysis. These regions are (1) regions in the field of view that were obstructed by the cylinders and (2) areas of the flow field that were poorly illuminated or cast into shadow.

Acquisition of the velocity field depends on first estimating the displacement field. This is achieved when the PIV images are interrogated using imaging software (Dantec DynamicStudio V1.45). The PIV images are divided into subregions referred to as interrogation windows that are 32 × 32 pixels in size and overlapped by 50%. The spatial resolution is 1.6 mm or $0.107D$ and the velocity map contains 127 × 127 vectors. The interrogation window offset technique was employed. This used the results from prior interrogation and validation runs to determine the offset amount of each interrogation window. Each window was offset by the discrete mean particle displacement (units: pixel) measured in a specific interrogation window. Window offsetting improves the strength of the displacement correlation peak relative to the random correlation peaks, which improved the reliability of the velocity field estimate. This process was repeated until the estimated velocity field was converged.

A number of validation procedures were implemented on the raw velocity vector maps to remove outliers:

1. Peak detection validation removes invalid velocity vector estimates. Vectors are rejected when the amplitude of the displacement correlation peak (signal) is less than 1.2 times the amplitude of the noise correlation peak. In addition, the correlation peak width must be within the range of 2–7 pixel units.
2. Range validation was then implemented to remove velocity vectors with a magnitude that was greater than a predetermined maximum value (ie, 12.0 m/s).
3. Moving average validation then followed where the averaging area consisted of 3 × 3 interrogation windows and the acceptance factor was 0.1 carried out over three iterations. Less than 1% of the vectors were rejected after image masking eliminated regions that were not suitable for interrogation due to poor illumination, shadows, and obstructions in the field of view.

The evaluation of the vector field between two successive frames yields an instantaneous vector map. The time-averaged velocity field for a specific location in the flow field is then evaluated over an ensemble of 80 instantaneous vector maps sampled at a frequency of 5 Hz.

D. Data Reduction Parameters and Measurement Uncertainties

The Reynolds number based on the single-cylinder diameter D and the upstream mean flow velocity U_∞ is defined as:

$$\mathrm{Re}_D = \frac{\rho_\infty U_\infty D}{\mu_\infty} \tag{5.2}$$

where ρ_∞ and μ_∞ are the density and dynamic viscosity of air, respectively. The Reynolds number was fixed at 6000 (subcritical flow regime).

The experimental uncertainty of the Reynolds number and axial flow velocity were estimated using a method reported by Holman and Gajda (1978) based on 20:1 odds and were within 1.1%. The uncertainty of the instantaneous PIV velocity measurements is primarily related to the estimation of the average particle displacement within an interrogation area. The uncertainties regarding the timing of the light sheet pulses, camera synchronization, and particle lag are not considered as significant sources of error. The determination of measurement uncertainty relating to particle displacement has been quantified analytically and by the generation of synthetic images with known parameter values (Westerweel, 1997, 2000, 2008). In order to provide a reasonable estimate of measurement error, the results given in Westerweel (2000) are considered to be applicable to the PIV algorithm used in this investigation. The displacement measurement error based on the mean particle image diameter of 2–3 pixels is 0.05 pixels. Therefore, the full-scale relative measurement error is 0.6%.

E. Interpretations of Measurement Data

A comparison of the raw PIV images using the external and the internal isotropic-planar illumination configurations is shown in Fig. 5.17. For both cases the CCD camera is focused at the gap region between the third (central) cylinder and the fourth (lower) cylinder. The camera's object plane that is coincident with the laser light sheet is at the midspan of the cylinders. The camera lens is a conventional, nontelecentric type. Consequently, due to perspective (parallax) effects the ends of the cylinders that are closest to the camera will obstruct the field of view. The cylinders will also appear to converge to a single point in the background of the image. The obstruction prevents seeding particles close to the cylinder walls from being captured in the PIV images. Unfortunately, this problem is unavoidable with a conventional (nontelecentric) lens, but efforts were made to minimize the parallax effect. The CCD camera was positioned as far from the object plane as possible without deteriorating the focus of the particle images in the object plane/light sheet. However, a level of compromise was necessary and the obstructed regions were masked before interrogation. Therefore, no velocity field information could be obtained from these regions.

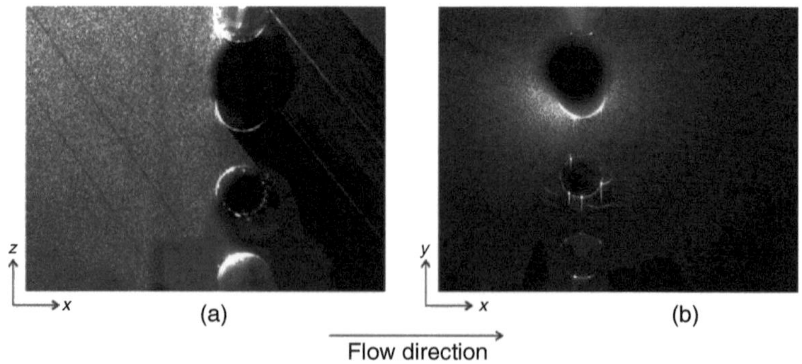

Figure 5.17 ***Raw PIV images of the flow around five side-by-side circular cylinders where all the cylinders are transparent.*** (a) External illumination; (b) internal isotropic illumination where the central cylinder houses the laser optics and radiates laser illumination outwards from the cylinder center (Atkins and Kim, 2015).

Fig. 5.17a shows a raw PIV image from the external illumination configuration. Multiple shadow regions are created by the cylinders, despite the cylinders being made from transparent acrylic tube. The illumination transmitted through cylinders created caustic patterns in the shadow that appear as "streak" lines.

Isotropic-planar illumination is shown in Fig. 5.17b where the laser light radiates outwards from the central cylinder with 360° coverage. The thickness of the light disc does not change significantly throughout the field of view. Therefore, the variation of the light intensity in the radial direction is roughly inversely proportional to the distance away from the center of the cylinder. The seeding particles in the flow also attenuate the transmission of the light due to scattering.

To illuminate the outer regions of the field of view, a relatively high light intensity is necessary. Consequently, regions close to the cylinder surface would have high levels of exposure that may cause localized blooming on the PIV image as shown from Fig. 5.17b. At a certain distance from the central cylinder, the light intensity will become too low to adequately illuminate the seeding particles. With insufficient illumination, the reliability of the velocity estimates will deteriorate as the signal to noise ratio will reduce.

Surface flaring and streaks in the light disc were also caused when seeding material accumulated on the cylinder surface. Occasionally, the deposited material increased light scattering which saturated the CCD detector (blooming). The rate of material accumulation was dependent on the experimental conditions and seeding concentration. This limited the total time of the measurement run and the cylinder had to be wiped clean when the buildup became too severe. Measurement runs ranged from 16 to 30 s.

Implementing internal isotropic-planar illumination requires a balance between a large field of view and blooming or pixel saturation around the illuminating cylinder. With the current experimental setup, it was possible to obtain full illumination of the field of view with limited blooming near the central cylinder.

Fig. 5.18a shows the velocity field measured using external illumination. The gap and downstream regions provide no velocity field information, which is a direct consequence of having these regions cast into shadow. Because the laser was offset by 37° with respect to the transverse plane of the cylinders and the spacing ratio between the cylinders is relatively large (eg, $T/D = 1.7$), partial illumination of the downstream region occurs where "strips" of laser light appear. Incomplete illumination of the downstream region creates spurious vectors or outliers due to the inadequate number of seeding particle pairs appearing within the interrogation areas during the PIV interrogation process. However, due to the angular offset, the upstream velocity field ahead of the cylinders has been well resolved.

In contrast, Fig. 5.18b shows the velocity field measured using internal isotropic-planar illumination where the entire flow field around the

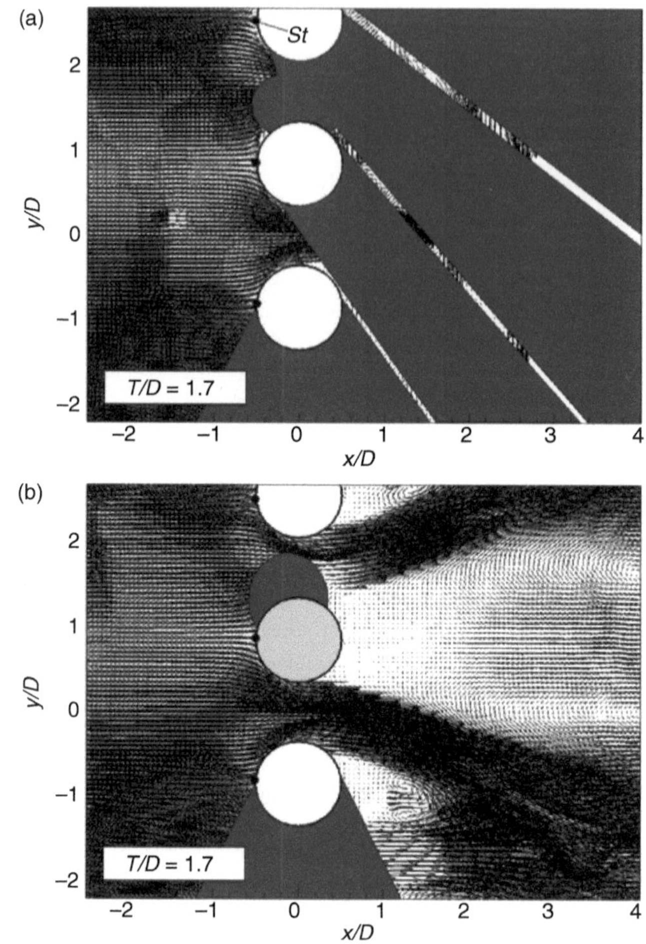

Figure 5.18 ***Time-averaged velocity vector field of the flow around five side-by-side circular cylinders at Re_D = 6000 where*** **T/D** ***= 1.7.*** (a) External illumination configuration; (b) internal isotropic-planar illumination, and the filled cylinder is the middle, illuminated cylinder (Atkins and Kim, 2015).

illuminated (middle) cylinder is revealed. The gap flow, the flow passing between the cylinders, that is, above and below the middle cylinder, is deflected towards the adjacent cylinders. Consequently, the near-wake structure behind the middle cylinder is wide whereas the near wake behind the adjacent cylinders is narrow.

The calculation of derived quantities based on the velocity vector field such as streamlines, reveals with a greater degree of clarity how the entire flow field around the cylinders develops. Referring to Fig. 5.19a (external illumination), only the upstream streamlines can be calculated due to the upstream

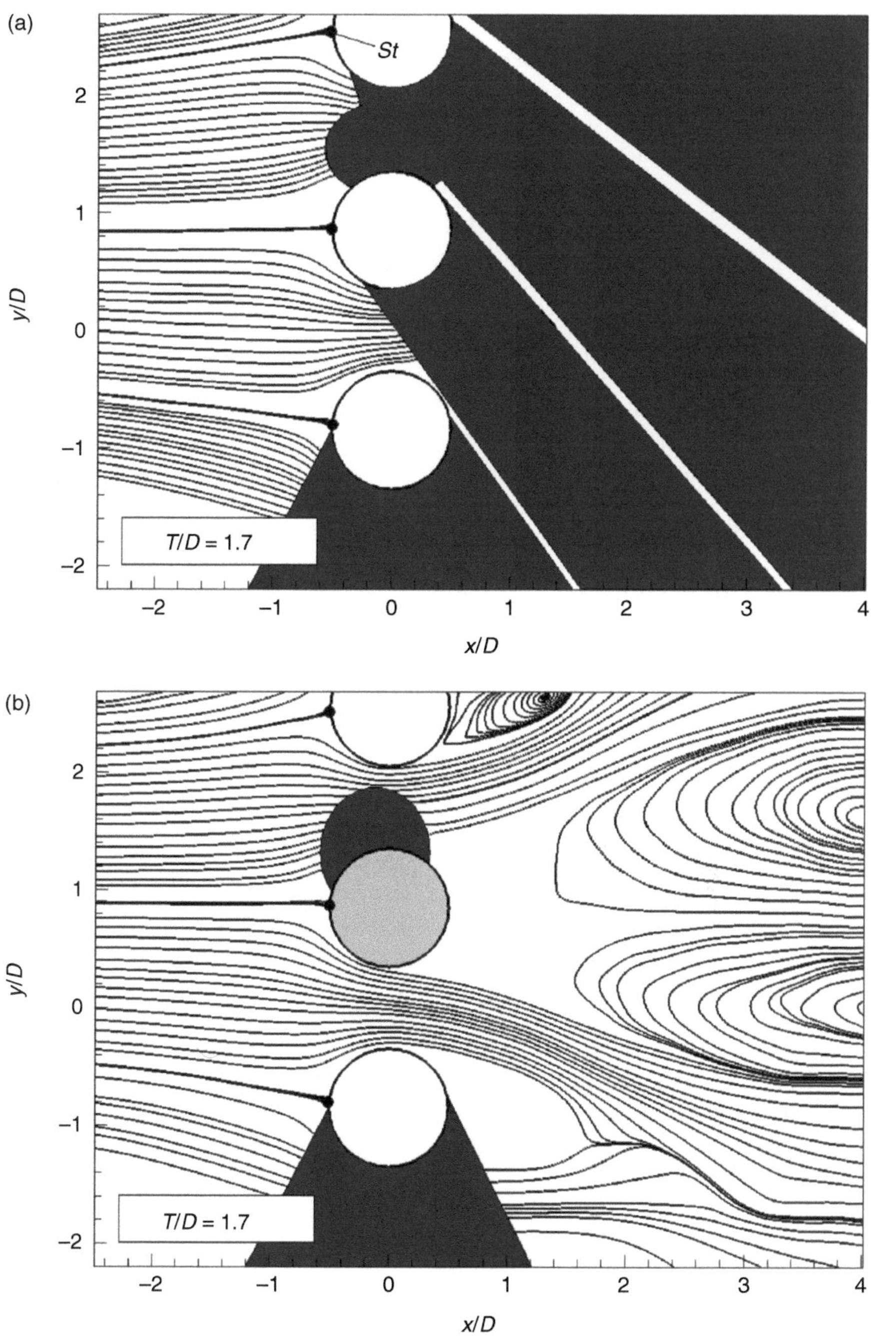

Figure 5.19 ***Time-averaged streamlines of the flow around five side-by-side circular cylinders at $Re_D = 6000$ where T/D = 1.7 and St denotes the stagnation point on the cylinder surface.*** (a) External illumination configuration; (b) internal isotropic-planar illumination (Atkins and Kim, 2015).

flow region being illuminated. No physical insight of the flow behavior in front of the cylinders can be linked to the gap and downstream wake behavior.

In contrast, streamlines measured using the internal isotropic-planar illumination in Fig. 5.19b demonstrate that the upstream streamlines *disperse* ahead of the cylinders. The dispersion indicates that the flow velocity ahead of the cylinder decreases relative to the free stream. From the flow images, it is also possible to determine the approximate location of the stagnation points on the cylinder surface. The stagnation points (*St*) are the points on the cylinders' surfaces where the oncoming flow is divided either side of the cylinders. Hence, as the approaching flow is divided, the streamlines bifurcate ahead of the cylinders. Furthermore, there is a high degree of flow acceleration in the gap region between the cylinders as the streamlines now bunch together when the flow passes between the cylinders.

Fig. 5.20 maps the strength of velocity fluctuations (V_{rms}) in the flow by calculating the RMS of the measured velocity field (Wernet, 2000). Hence, it also indicates turbulence intensity levels in the flow field if we assume that random fluctuations in the measurements are small in relation to the turbulence levels. From Fig. 5.20a, b, it can be seen that the V_{rms} values corresponding to the upstream flow region ahead of the cylinders are similar for the external and internal illumination configurations. However, the internal illumination setup (Fig. 5.20b) also reveals information about the gap flow and the downstream regions simultaneously. Again, no such information can be retrieved with external illumination (Fig. 5.20a).

The overall degree of fluctuation in the upstream and gap regions shown in Fig. 5.20b is quite low compared to the downstream region due to natural wake shedding behavior. It appears that the wide wake behind the central cylinders has far lower velocity fluctuation (V_{rms}) values compared to the narrower wake emanating from behind the adjacent cylinders. This result may support the hot-film measurements of Sumner et al. (1999) who reported that the wider wake has a lower shedding frequency compared to a narrower wake.

The overall flow pattern around the row of five side-by-side cylinders is symmetrical about the central cylinder (Fig. 5.21). The stagnation points on the adjacent cylinders, corresponding to the upstream flow, are inclined towards the middle cylinder by similar amounts whereas the stagnation point of the middle cylinder is aligned with the nominal free-stream flow. The gap flows in the wake of the adjacent cylinders are deflected away from the central cylinder, indicating a symmetric-biased flow pattern, which is similar to the case of three side-by-side circular cylinders reported by Sumner et al. (1999).

F. Conclusions

A new laser illumination configuration, "isotropic-planar" illumination, has been introduced. This new method illuminates regions of the flow field that were previously cast into shadow by creating a 360°disc of light that

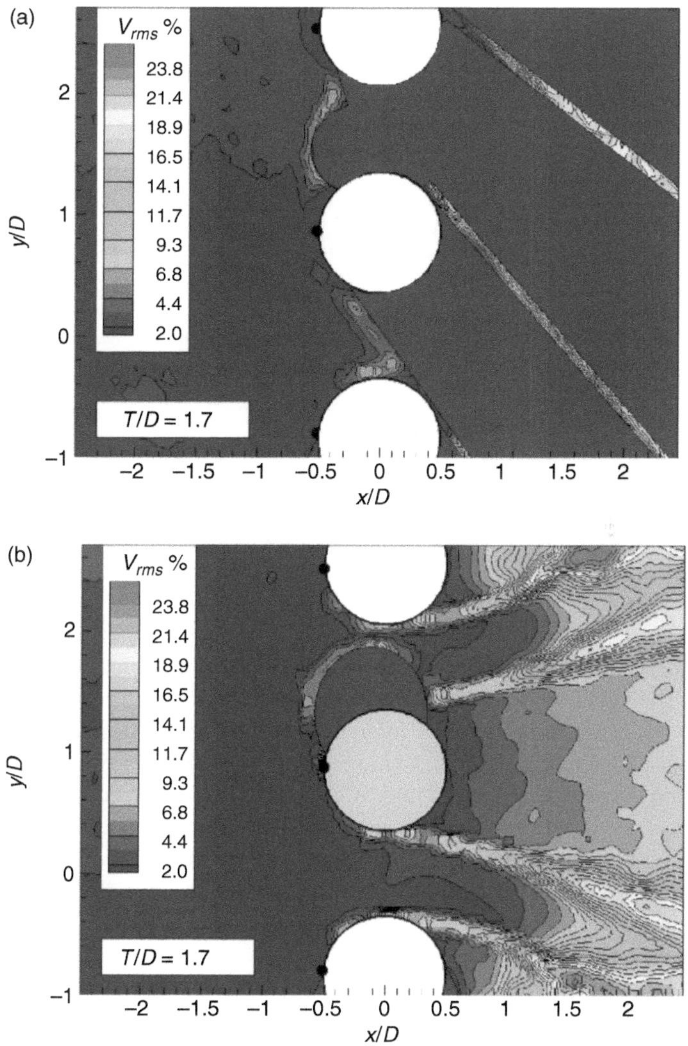

Figure 5.20 ***RMS map of the velocity field.*** (a) External illumination; (b) internal isotropic-planar illumination (Atkins and Kim, 2015).

is generated from a new optical arrangement, placed inside of a transparent test body. The test body is, therefore utilized as a means of introducing the laser light internally into the test section. This is in contrast to an externally introduced light sheet. A comparison between the conventional external and the internal isotropic-planar illumination configurations has been presented in this example, using five side-by-side circular cylinders in a crossflow plane.

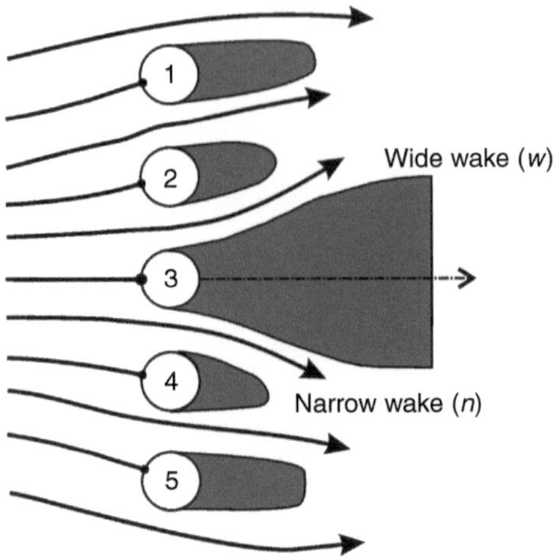

Figure 5.21 ***Schematic of symmetric-biased wake pattern behind five side-by-side circular cylinders showing that the stagnation points of the adjacent cylinders are inclined towards the central cylinder and the gap flows are biased away from the same cylinder*** (Atkins and Kim, 2015).

The results indicate that there is an overall symmetrical flow pattern about the central cylinder where the stagnation points of the adjacent cylinders are inclined towards the middle cylinder. The gap flows in the wake are then deflected away from the central cylinder. Such symmetrical flow behavior could indicate that there is a possible link between the upstream flow and the downstream wake where the gap flow has biased behavior.

The new illumination method may also be applied to study the velocity fields of other configurations, such as multirow tube bundles. Employing external illumination, the region directly between the rows of cylinders typically would have been obstructed by the surrounding cylinders. Internal isotropic-planar illumination offers an alternative method to overcome some of the limitations of conventional illumination methods.

REFERENCES

Adrian, R.J., 1991. Particle-imaging techniques for experimental fluid mechanics. Annu. Rev. Fluid Mech. 23, 261–304.

Adrian, R.J., 2005. Twenty years of particle image velocimetry. Exp. Fluids 39, 159–169.

Adrian, R.J., Westerweel, J., 2011. Particle Image Velocimetry. Cambridge University Press, Cambridge.

Atkins, M.D., Kim, T., 2015. Isotropic-planar illumination for PIV experiments. Exp. Fluids 56, 63.

Baker, C.J., 1979. The laminar horseshoe vortex. J. Fluid Mech. 95, 347–367.

Baker, C.J., 1980. The turbulent horseshoe vortex. J. Wind Eng. Ind. Aerod. 6, 9–23.

Ballio, F., Bettoni, C., Franzetti, S., 1998. A survey of time-averaged characteristics of laminar and turbulent horseshoe vortices. J. Fluid. Eng. – T. ASME 120, 233–242.
Bryanston-Cross, P.J., Towers, C.E., Judge, T.R., Towers, D.P., Harasgama, S.P., Hopwood, S.T., 1992. The application of particle image velocimetry (PIV) in a short-duration transonic annular turbine cascade. J. Turbomach. 114, 504–509.
Dantec Dynamics A/S, 2002. 80X70 Light-sheet Series Installation & User's Guide, Publication no. 9040U6052 (1/07/2002).
Holman, J.P., Gajda, Jr., W.J., 1978. Experimental Methods for Engineers. McGraw-Hill, New York.
Hung, C.M., Sung, C.H., Chen, C.L., 1991. Computation of saddle point of attachment, AIAA Paper 91-1713.
Hunt, J.C.R., Abell, C.J., Petetka, J.A., Woo, H., 1978. Kinematical studies of the flows around free or surface-mounted obstacles; applying topology to flow visualization. J. Fluid Mech. 86, 179–200.
Ishigai, S., Nishikawa, E., Nishimura, K., Cho, K., 1975. Experimental study of structure of gas flow in tube banks with tube axes normal to flow, Part II on the structure of gas flow in single column, single row, and double row tube banks. B. JSME 18 (119), 528–535.
Keane, R.D., Adrian, R.J., 1992. Theory of cross-correlation analysis of PIV images. Appl. Sci. Res. 49, 191–215.
Khan, M.J., Ahmed, A., 2005. Topological model of flow regimes in the plane of symmetry of a surface-mounted obstacle. Phys. Fluids 17, 045101.
Khan, M.J., Ahmed, A., Trosper, J.R., 1995. Dynamics of the juncture vortex. AIAA J. 33, 1273–1278.
Koschatzky, V., Moore, P.D., Westerweel, J., Scarano, F., Boersma, B.J., 2011. High speed PIV applied to aerodynamic noise investigation. Exp. Fluids 50, 863–876.
Moretti, P.M., Cheng, M., 1988. Experimental study of the flow field downstream of a single tube row. Exp. Therm. Fluid Sci. 1, 69–74.
Perry, A.E., Chong, M.S., 2003. Interpretation of flow visualization. In: Smits, A.J., Lim, T.T. (Eds.), Flow visualization, Imperial College Press, London, pp. 1–26.
Prasad, A.K., Adrain, R.J., Landreth, C.C., Offutt, P.W., 1992. Effect of resolution on the speed and accuracy of particle image velocimetry interrogation. Exp. Fluids 13, 105–116.
Raffel, M., Willert, C.E., Kompenhans, J., 1998. Particle Image Velocimetry: A Practical Guide. Springer, Heidelberg.
Seal, C.V., Smith, C.R., Rockwell, D., 1997. Dynamics of the vorticity distribution in endwall junctions. AIAA J. 35, 1041–1047.
Simpson, R.L., 2001. Junction flows. Annu. Rev. Fluid Mech. 33, 415–443.
Sumner, D., Wong, S.S.T., Price, S.J., Paidoussis, M.P., 1999. Fluid behavior of side-by side circular cylinders in steady cross-flow. J. Fluid. Struct. 13, 309–338.
Tobak, M., Peake, D.J., 1979. Topology of Two-dimensional and Three-dimensional Separated Flows, presented at the AIAA 12th Fluid and Plasma Dynamics Conference, AIAA paper 79-1480.
Visbal, M.R., 1991. Structure of laminar juncture flows. AIAA J. 29, 1273–1282.
Wang, Z.J., Zhou, Y., 2005. Vortex interactions in a two side-by-side cylinder near-wake. Int. J. Heat Fluid Fl. 26, 362–377.
Wernet, M.P., 2000. Application of DPIV to study both steady state and transient turbomachinery flows. Opt. Laser Technol. 32, 497–525.
Wernet, M.P., Pline, A., 1993. Particle displacement tracking technique and Cramer-Rao lower bound error in centroid estimates from CCD imagery. Exp. Fluids 15, 295–307.
Westerweel, J., 1997. Fundamentals of digital particle image velocimetry. Meas. Sci. Technol. 8, 1379–1392.
Westerweel, J., 2000. Theoretical analysis of the measurement precision in particle image velocimetry. Exp. Fluids 29 (Suppl. 1), S003–S012.

Westerweel, J., 2008. On velocity gradients in PIV interrogation. Exp. Fluids 44, 831–842.
Westerweel, J., Dabiri, D., Gharib, M., 1997. The effect of a discrete window offset on the accuracy of cross-correlation analysis of digital PIV recordings. Exp. Fluids 23, 20–28.
Willert, C.E., Gharib, M., 1991. Digital particle image velocimetry. Exp. Fluids 10, 181–193.
Wilson, B.M., Smith, B.L., 2013. Uncertainty on PIV mean and fluctuating velocity due to bias and random errors. Meas. Sci. Technol. 24, 035302.
Xu, S.J., Zhou, Y., So, R.M.C., 2003. Reynolds number effect on the flow structure behind two side-by-side cylinders. Phys. Fluids 15, 1214–1219.

PART 2

Measurements in Heat Transfer

CHAPTER 6

Point Temperature Measurements

D.J. Rossouw
Nuclear Science Division, School of Mechanical Engineering, University of the Witwatersrand, Johannesburg, South Africa

6.1 INTRODUCTION

Prior to introducing techniques for "temperature" measurement, it is instructive to begin by defining what temperature is. Temperature, in a qualitative manner, determines the degree of warmth or coldness felt and is a relative indicator showing the level of energy. Therefore, heat transfer is how heat (or energy) is transported from a high level of energy (or temperature) to a low level until an equilibrium state between the two is reached (Beverly, 1995).

Although numerous efforts have been devoted to quantify and define temperature in absolute and relative manners, the Celsius scale (denoted typically as °C) has been widely used. The Celsius (also called "centigrade") scale is based on taking the difference in temperature between the melting point of ice and the boiling point of water and dividing this into 100 equal increments. In the scientific community, the absolute scale of the Celsius scale from "a naturally defined zero" is commonly used, which is defined from ideal gas relations, that is, the temperature is zero when the pressure of the ideal gas is zero. This is called the "Kelvin scale." The temperature in the Kelvin scale is related to that in the Celsius scale as $B = A + 273.16$ where A is temperature in °C scale and B is the converted temperature in K scale (no degree symbol used).

This chapter reviews some commonly used temperature/heat flux measurement techniques including the methods used for calibrations. As in flow measurements, either the temperature at a physical point or the temperature field (or map) is of interest. Measurement techniques for temperature at a point are covered in this chapter, whilst the measurement techniques for temperature field (eg, thermochromic liquid crystals and infrared thermograph) are discussed in Chapters 7 and 8, separately.

Application of Thermo-Fluidic Measurement Techniques
http://dx.doi.org/10.1016/B978-0-12-809731-1.00006-X

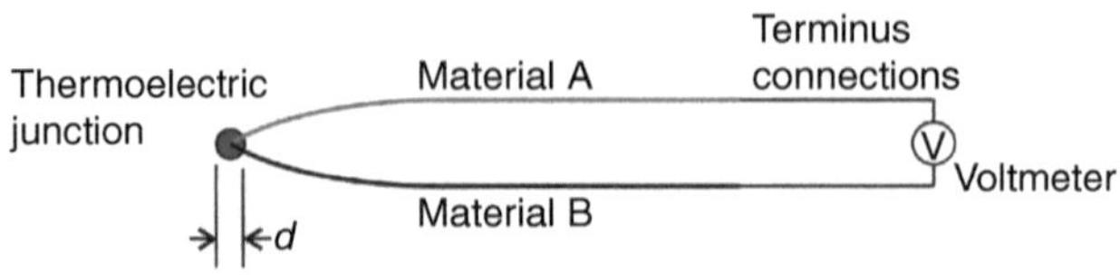

Figure 6.1 *Typical thermocouple circuit.*

6.2 THERMOCOUPLES

6.2.1 Background

Thermocouples are one of the commonly used thermometers that can be used to quantify temperature. The widespread use of thermocouples in temperature measurement is associated with their low cost, robustness, and wide measurable temperature range. As shown in Fig. 6.1, a thermocouple consists of two dissimilar wires. When the temperature of the two junctions is different, a small current flows through the circuit, as first discovered by Seebeck in 1823. This current is proportional to the temperature difference between the two junctions, which is often called the "Seebeck effect" [see Childs (2001) for details].

Many types of thermocouples have been developed, which are classified depending on different combinations of the two dissimilar wires (Table 6.1). The selection of a particular thermocouple depends on the required temperature range, cost, chemical stability, material compatibility, and mechanical limitations.

Table 6.1 Standardized Thermocouples

Type	Temperature range [°C]	Materials
T	−262 ~ +850	Copper/copper–nickel alloy (constantan)
J	−196 ~ +700	Iron/constantan
E	−268 ~ +800	Nickel–chromium alloy (chromel)/ constantan
K	−250 ~ +1100	Chromel/nickel–aluminum alloy (alumel)
N	0 ~ +1250	Nickel–chromium silicon/nickel–silicon–magnesium alloy
B	+100 ~ +1750	Platinum–30% rhodium/platinum–60% rhodium
S	0 ~ +1500	Platinum–10% rhodium/platinum
R	0 ~ +1600	Platinum–13% rhodium/platinum

Source: Childs (2001).

6.2.2 Size of the Junction

The junction of the two metal wires typically forms a bead (sphere) as a result of welding. This may sometimes cause problems, for example, if flush surface installation is required. The size of the bead plays an important role in determining the response time of a thermocouple due to thermal capacity: the smaller the bead size, the quicker becomes the response to the change of temperature field in the vicinity of the thermocouple. Thin bead thermocouples have been developed (Omega™) which are very fragile, but have a very quick response time and are effective when flush mounting is required.

6.2.3 Calibration and Data Acquisition

It is not always necessary to calibrate thermocouples. The use of standardized thermocouples such as those listed in Table 6.1 allows the user to rely on the tolerance stated by the manufacturer and use standard tables to convert the signal from the thermocouple into temperature. State-of-the-art equipment contains built-in calibration mechanisms that allow the user to easily connect a thermocouple to the device and read temperature directly.

There are two main types of conventional thermocouple calibration methods that exist. These include the standard ice-bath method and cold-junction compensation (CJC) (National Instruments, 2011). The built-in CJC method has gained popularity since the ice-bath method is more tedious to use and perhaps less accurate. The ice-bath method is useful for thermocouples that do not require a high degree of accuracy.

Fig. 6.2 schematically shows a water bath with a controllable water temperature for thermocouple calibration, which makes use of naturally known temperatures such as the triple point of water (0.01°C) or the boiling

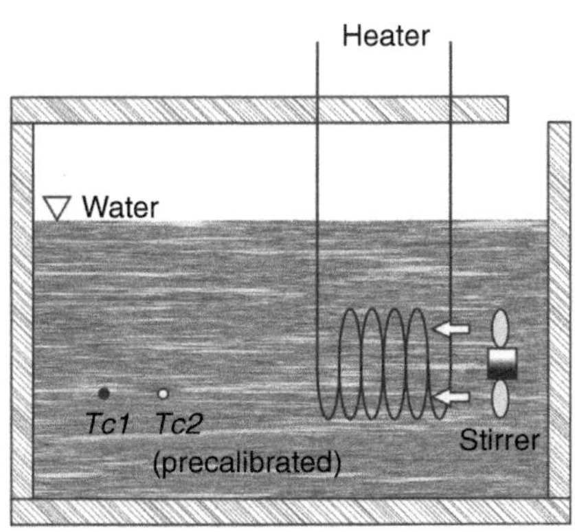

Figure 6.2 ***Schematic of thermocouple calibration setup with a precalibrated thermocouple (T_{C2}) in a controllable water temperature bath.***

temperature of water (100°C) in ambient pressure. The readout from a pre-calibrated thermocouple (T_{C2}) is used to calibrate the target thermocouple (T_{C1}). The readout signals from both thermocouples are monitored by electronic devices (eg, a multimeter).

The other conventional method which is commonly used is the CJC method. Much like the ice-bath method, the temperature of that which is being measured requires a reference temperature (0.01°C in the case of the ice-bath). Such equipment making use of the CJC method uses electronic temperature sensors, consisting of, for example, thermistors and thermostats, in order to calculate this reference temperature. The most common electrical devices used are the resistance temperature detector (RTD) and also silicon-based temperature sensors (Maxim Integrated, 2007). The reference temperature is measured and then applied to the thermocouple being used, and the measured temperature is then displayed as a result.

A RTD works on the basis that the resistance changes in a material as the temperature in the material changes. A RTD made from a known material, usually copper, nickel, or platinum, shows resistance at a particular temperature. Following a temperature change, there is a change in resistance and the temperature of the measured reference point can be determined based on the materials' resistance–temperature relationship. Platinum is most utilized in a RTD as it has maximum accuracy due to the material's high response time. Also, it has a wider range of temperature operation.

The other common electrical device is the silicon-based temperature sensor known as a silicon bandgap temperature sensor. This is a silicon-based sensor that is embedded into the electrical circuit of the digital instrument. The device measures the voltage difference, using diodes and transistors, across the band gap and then converts it into a temperature based on the band gap's voltage–temperature relationship. This is a cheaper way of integrating reference temperature measurement into state-of-the-art equipment rather than using RTDs.

6.2.4 Thermocouple Connection Leads

There are many different ways to connect thermocouples. However, care should be taken so that system errors are minimized. If it is required to extend the wires to the thermocouple, the same material must be connected to both junctions, so as to minimize the additional voltage difference between the two thermocouple junctions. Ensure that solid contact is made between the measurement device connection and the wires of the thermocouple junctions.

6.2.5 Types of Measurement Devices

There are many types of electronic instruments that are used to measure temperature. It can be determined by level of accuracy, cost, and ease of use. If thermocouples are to be used, it is suggested that the device has a good resolution in the range of the thermocouples' resolution. Also, the device being used should make use of one of the built-in calibration methods already discussed to minimize the need to perform any type of calibration.

Fig. 6.3a shows an image of a thermometer, a Fluke™ 52 II with two T-type thermocouples attached. This device is effective for stable temperature measurements with resolution of 0.1°C. The user selects the type of thermocouple attached and then reads the measured temperature. Fig. 6.3b shows a multifunctional data acquisition unit, an Agilent™ LXI switch unit.

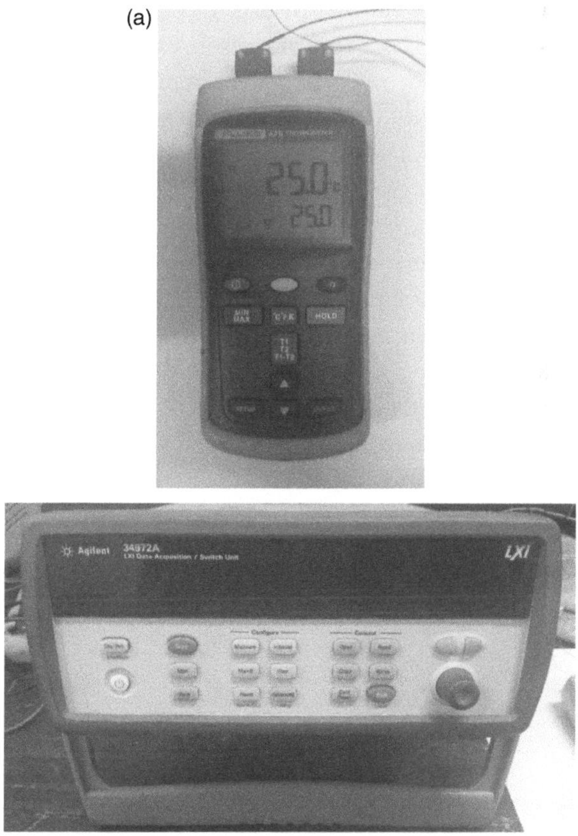

Figure 6.3 ***Temperature reading devices.*** (a) Fluke™ 52 II; (b) Agilent™ 34972A switch unit.

This device is suitable for fast temperature readings with a function that can store multiple readings and then display the average value. The device has a resolution of 0.1°C.

6.3 SURFACE-MOUNTABLE THERMOCOUPLES

Many engineering measurements are involved with surface temperature measurements. When the surface of a finite-thickness solid experiences fluid flow normal/parallel to it, a bead-type thermocouple is commonly used to measure its temperature. A small hole is drilled from the backside of the solid until the thermocouple is flush with the surface, which is then potted with thermal epoxy. This allows the thermocouple to slightly protrude into the channel as illustrated in Fig. 6.4a. Such protrusion may change the velocity of flow in the boundary layer and hence detrimentally alter the local condition for heat transfer on the surface.

Alternatively, the tip of the thermocouple can be removed so that there is no protrusion of the thermocouple bead as illustrated in Fig. 6.4b. In this case, an epoxy (conductive) filler has to be used to fill the gap caused by the spherical thermocouple bead. The mismatch of thermal conductivity between the filler and the thermocouple bead may increase the conduction resistance, causing readout errors of the thermocouple.

A variety of different flush mountable film-type thermocouples (Fig. 6.4c) can be used instead and are currently available (Fig. 6.5a, b). The attachment of these thermocouples (approximately 20 μm thick) on the surface can reduce the flow disturbance caused by the physical presence of bead-type thermocouples. However, the use of aerosol-type glue for thermocouple attachments may increase the thermal resistance between the surface and the film thermocouple. The thermocouple in Fig. 6.5a (10 μm thick) is thinner than the thermocouple in Fig. 6.5b (20 μm thick). The

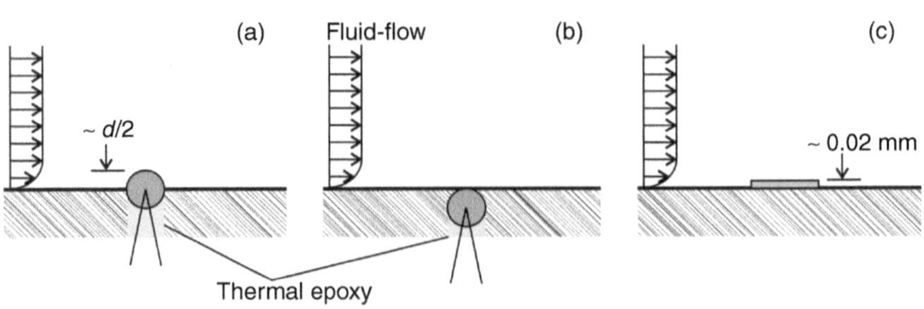

Figure 6.4 ***Examples for installing a thermocouple to measure surface temperature.***

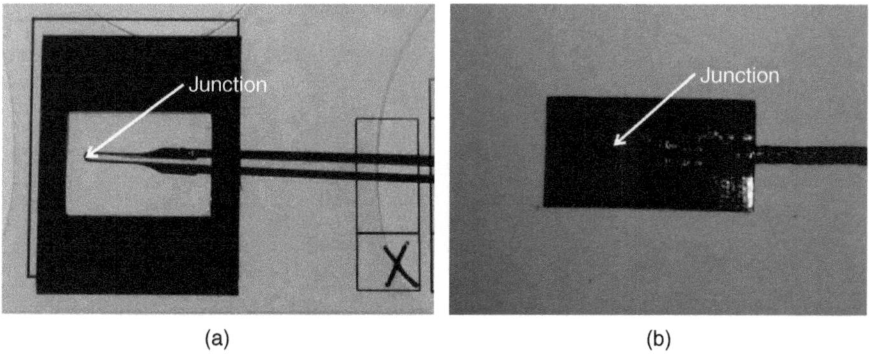

Figure 6.5 ***Thermocouple types (Omega™).*** (a) Thin film T-type thermocouple; (b) laminated thin T-type thermocouple.

former has a slight advantage in minimizing flow disruption over a surface but electrical insulation is required if mounted on a metallic surface. Furthermore, it has a much faster response time than the latter and is also advantageous in terms of high-speed measurements and better accuracy. Extreme care must be taken due to its thinness and fragile build, avoiding bending the thermocouple and avoiding contact with the junction tip. The lamination in the latter (Fig. 6.5b) is intended for ease of surface mounting when required (Omega Engineering, inpress).

Example 6.1: The Thermal Interaction of a Buoyant Plume From a Calandria Tube With an Oblique Jet

A. Background

The CANDU-6 reactor core contains an array of 380 calandria tubes (CTs) which are surrounded by heavy water (D_2O). Inside these CTs are concentric pressure tubes (PTs) that contain modular nuclear fuel bundles. The annular space between the calandria tube and a pressure tube is filled with CO_2 gas that is a thermal insulating medium between the two concentric tubes. Under normal operating conditions there is a continuous flow of coolant through the pressure tube/fuel bundle; however, abnormal conditions may result in a failure of the coolant system and compromised cooling capability required for the fuel (Fig. 6.6). This scenario is referred to as a loss-of-coolant accident (LOCA) (IAEA, 2000).

With the LOCA, there is significant local heating which causes localized deformation of the pressure tube, where "sagging" or "ballooning" occurs depending on the internal pressure. The pressure tube deformation progresses until it makes contact with the inner surface of the calandria tube. The contact allows conduction heat transfer to take place across the

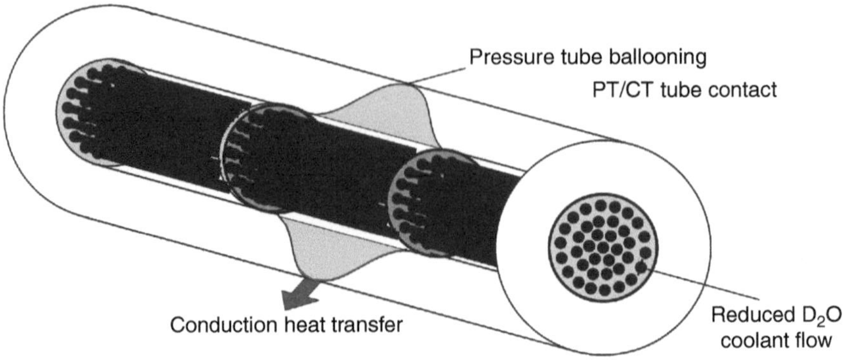

Figure 6.6 ***A loss-of-coolant accident (LOCA) where there is ballooning type deformation of the pressure tube.***

tube walls and an associated spike of the heat flux into the moderator. The integrity of both the CT and PT is exposed under progressive heating.

To prevent the exacerbation of the LOCA, a safety feature is to circulate the moderator (D_2O) along the inner periphery of the calandria tank and subsequently to create a reverse-flow jet directed towards the calandria tubes experiencing the LOCA (Fig. 6.7). Previous studies (Nandan et al., 2010; Nguyena et al., 2008; Dimmick et al., 2002; Kim et al., 2006) have shown that the external and additional heat removal by manipulating the circulating moderator can suppress heating on the outer surface of the CTs suffering from the LOCA.

The external D_2O flow, in the form of the reverse-flow jet which approaches the CTs, interacts with rising natural convection (ie, buoyant plumes) that occurs (Incropera and DeWitt, 1990). This study, in turn, aims to understand how buoyant plumes developed from the CTs suffering from the LOCA are altered by the approaching reverse-flow jet in a simplified configuration: a circular heat source (simulating a hot calandria tube) cooled by an air impinging jet. The intensity of the buoyant plume and the reverse-flow jet is measured by a nondimensional number called the Archimedes number. Particular focus is placed on how the oblique positioning of the reverse-flow jet (from cocurrent to countercurrent configuration) with respect to rising natural convective flow influences overall natural convective cooling characteristics on a single calandria tube for an arbitrary Archimedes number.

B. Experimental Details

B1. Test Rig and Instrumentation

For heat removal by round jet impingement, air at ambient conditions drawn by a centrifugal fan was discharged from a round jet nozzle. The mass flow rate of the round jet was adjusted to fix the jet Reynolds number Re_j, based upon the round jet diameter D_j, at 1500 to provide a set Archimedes number

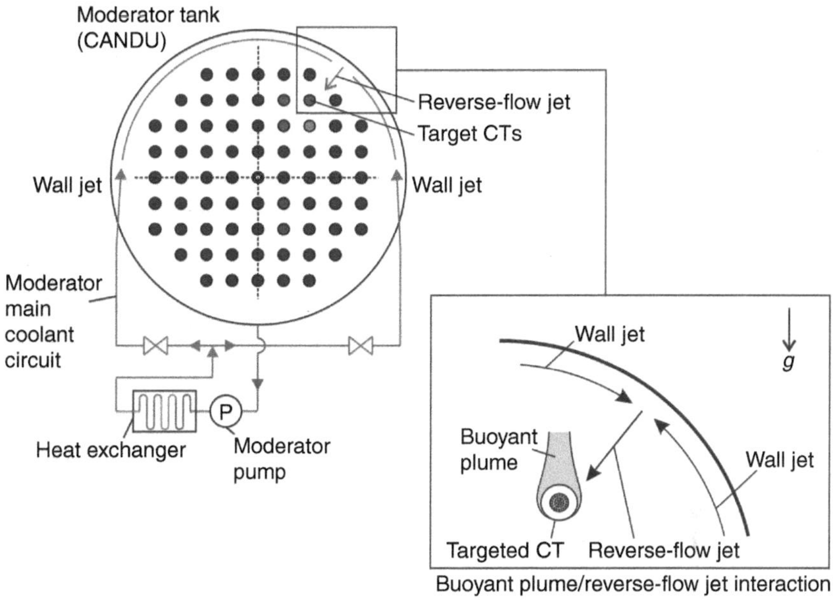

Figure 6.7 ***Schematics of counter wall jets that collide and form a reverse-flow jet on the periphery of the moderator tank. The reverse-flow jet then interacts with a buoyant plume caused by the heated calandria tube.***

Ar_D of 0.1. The round jet nozzle was mounted on a linear traverse system so that the jet exit to target cylindrical element distance, H/D_j, could be systematically varied. The target cylinder to round jet diameter ratio, D/D_j, was fixed to be 2.7. Other test setup details and conditions are listed in Table 6.2.

To characterize the present round jet, jet centerline velocity was traversed using a single component hot-wire anemometer (Dantec™) mounted on a linear traverse system along the z-axis (Fig. 6.8a). Prior to measurement, the hot-wire anemometer was calibrated using a Pitot tube whose pressure readings were recorded by a differential pressure transducer (DSA™, Scanivalve Inc.).

Table 6.2 Parameters of Test Setup and Conditions

Test setup parameter	Value
Cylinder (tube) diameter, D	0.070 m
Round jet diameter, D_j	0.026 m ($D/D_j = 2.7$)
Archimedes number, Ar_D	(based on $\Delta T = T_s - T_{amb} = 45.5$ K
where $\beta = 0.0032$ 1/K evaluated at T_f	at natural convection setting)
Prandtl number, Pr	0.7
Rayleigh number, Ra_D	1.15×10^6
Reynolds number, Re_j	1500 ($w_e \approx 1.0$ m/s)
Constant heat flux, q''	405 W/m^2

To impose constant heat flux on the cylindrical element that simulates the single calandria tube representing a hot reactor core region, a film-type heating element (Omega™) with a thickness of 0.40 mm was used. Double-sided adhesive tape was inserted between the heating element and the cylinder surface (Fig. 6.8b). To directly measure the heat flux removed from the heating element by round jet impingement, a film heat flux gauge (Omega™), 0.30 mm thick, was used. The heat loss passing through the acrylic tube was also measured using a heat flux sensor in order to obtain the actual heat flux on the surface of the cylinder. To measure local temperature along the cylinder circumference, a foil K-type thermocouple (Omega™), 0.01 mm in thickness, was flush-mounted on the cylinder surface using spray epoxy at a cylinder midspan, coinciding with the jet axis. A temperature scanner (Agilent™) was used to obtain the temperature readings used in finding the heat transfer (Nusselt number). The cylinder was rotated with a 5.0° increment from $\lambda = 0°$ (stagnation point always coinciding with the centerline axis of the jet) covering the entire cylinder circumference. Note that no flow disturbance or change by the surface-mounted foil thermocouples was assumed due to their thinness. A bead T-type thermocouple (Omega™) was used to monitor the jet temperature (T_j).

B2. Data Reduction Parameters and Measurement Uncertainties

Multiple studies (Misumi et al., 2003; Misumi et al., 1999; Wang et al., 1990) concerning natural convection have used the same definition based on the tube diameter of the cylinder. Therefore, for natural convection in this present study, the Grashof number (the ratio of the buoyancy forces and viscous forces) is defined as:

$$\mathrm{Gr}_D = \frac{g\beta\Delta T D^3}{\nu^2} \tag{6.1}$$

where g is the acceleration due to gravity, β is the fluid volumetric expansion coefficient, ΔT is the representative temperature difference between the cylinder surface and the ambient air at rest ($T_s - T_{amb}$), D is the tube diameter, and ν is the kinematic viscosity. Then, the Rayleigh number (a product of the Grashof and Prandtl numbers) is calculated as $\mathrm{Ra}_D = \mathrm{Gr}_D\mathrm{Pr}$. For consistency with other studies, the properties of the fluid are evaluated at the film temperature, T_f:

$$T_f = \frac{T_s + T_{amb}}{2} \tag{6.2}$$

where T_s is the cylinder surface temperature and T_{amb} is the ambient air temperature at rest.

The Archimedes number, which indicates the overall significance of buoyancy in mixed convection (a combination of forced and natural convection) flows, is defined as:

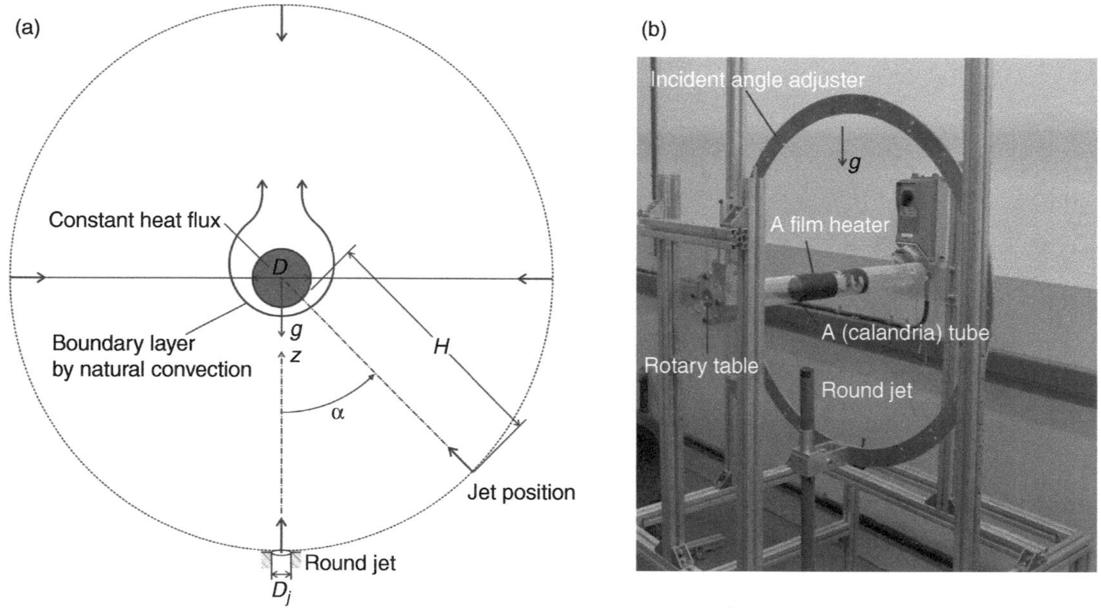

Figure 6.8 ***Test setup.*** (a) Schematic of the test configuration where α is the jet's oblique angle measured from the g-axis; (b) a photograph of the test rig.

$$\mathrm{Ar}_D = \frac{g\beta\Delta TD}{w_e^2} \tag{6.3}$$

where w_e is the centerline convective flow velocity at the exit. In this study, Ar_D is set at 0.1 as an arbitrary choice.

Based on the exit velocity w_e that satisfies the Archimedes number of 0.1, the jet Reynolds number is defined as a function of the jet diameter D_j as:

$$\mathrm{Re}_j = \frac{w_e D_j}{\nu} \tag{6.4}$$

During the entire experiment, the jet Reynolds number was fixed at 1500.

Heat transfer characteristics were evaluated using the Nusselt number defined, based on the tube diameter D as a function of the azimuth angle, as:

$$\mathrm{Nu}_D(\lambda) = \frac{h(\lambda)}{k_f/D} \tag{6.5}$$

where

$$h(\lambda) = \frac{q''}{T_s(\lambda) - T_j} \tag{6.6}$$

Here, q'' is the net heat flux emitted from the tube surface measured directly by a film heat flux gauge and heat loss was accounted for, k_f is the thermal conductivity of air, determined at the film temperature, whilst $T_s(\lambda)$ and T_j are the temperatures measured along the cylinder circumference at an arbitrary azimuth angle λ and at the jet exit $z = 0$, respectively. The origin of the azimuth angle λ is varied dependent on the jet's oblique angle.

The uncertainty associated with the azimuth angle was found to be within $\pm$ 0.5°. The measurement uncertainty of the Reynolds number was estimated using a method reported in Coleman and Steele (1999) (based on 20:1 odds) and found to be within 1.2%. The resolution of the temperature readings from the temperature scanner was found to be $\pm$ 0.1 K for each thermocouple. For heat flux measurement, the error associated with the voltmeter scanner was estimated to be less than $\pm$ 5 μV, resulting in an error less than $\pm$ 6 W/m^2 for heat flux (details to be discussed in section 6.4). Therefore, the uncertainty of the Nusselt number was calculated to be within 4.5%.

C. Discussion of Results

C1. Natural Convection (Buoyant Plume) from a Horizontal Circular Tube

Natural convection from a horizontally positioned circular cylinder serves as a fundamental configuration with many direct engineering applications.

Therefore, a multitude of studies have been reported (Misumi et al., 2003; Misumi et al., 1999; Wang et al., 1990). A net buoyancy force as a result of body forces (gravitational forces) and a present density gradient causes the motion of the fluid flow around the cylinder (Incropera and DeWitt, 1990).

At $\lambda = 0°$ (see the inset of Fig. 6.9), viscous forces are dominant. These viscous forces cause the fluid to form a boundary layer around the cylinder as illustrated in Fig. 6.8. The boundary layer thickness tends to increase as the angle λ increases (or moving away from $\lambda = 0°$), developing the boundary layer (Incropera and DeWitt, 1990).

To provide credibility for the present test setup and data, natural convection from a horizontal long cylinder emitting a constant heat flux has been characterized for $Ra_D = 1.15 \times 10^6$. Fig. 6.9 exhibits the distribution of the local Nusselt number along the circumference of a circular cylinder, showing a symmetric heat transfer distribution with respect to the gravitational axis, that is, the *g*-axis.

Heat transfer near $\lambda = 0°$ fluctuates but in general has a high value followed by a monotonic decrease until the base region, $\lambda = 180°$, whose value is found to increase slightly. The present Rayleigh number implies that the natural convective flow within the boundary layer remains laminar over the entire surface. Morgan (1975) has numerically shown that if the entire surface is covered by the laminar boundary layer, for example, $Ra_D < 10^9$, the distribution of the local heat transfer is characterized by a maximum at $\lambda = 0°$ and a monotonic decay with increasing λ as consistently observed in the present study.

A particular study performed by Atmane et al. (2003) focused on the natural convection around a horizontal cylinder with a wide range of Rayleigh numbers. Using their Rayleigh number matching $Ra_D = 1.15 \times 10^6$,

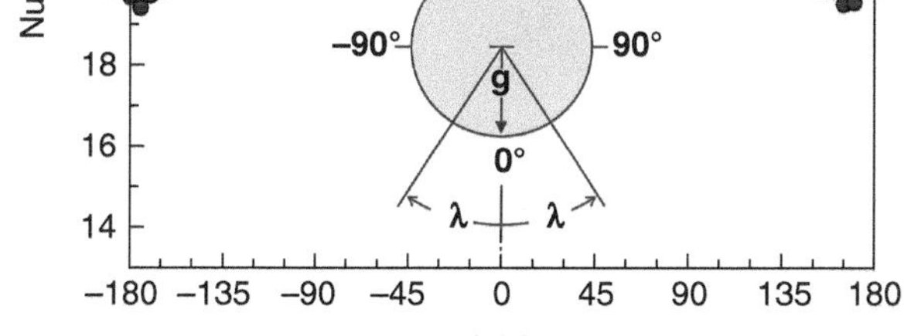

Figure 6.9 ***Local Nusselt number around a single circular tube emitting a constant heat flux (simulating a calandria tube) at $Ra_D = 1.15 \times 10^6$.***

and the area-averaged Nusselt numbers, an average Nusselt number of 20 was obtained. This is in close agreement with the area-averaged Nusselt number of 22.01 for the present data (within +10% difference).

C2. Cocurrent Jet-Buoyant Plume Interaction

Whilst natural convection removes heat from the tube, the jet is issued at an oblique angle of $\alpha = 0°$ as illustrated by the inset of Fig. 6.10. In this configuration, the direction of the jet is favorable to the buoyant plume from the tube – the cocurrent jet. The Archimedes number is set at $Ar_D = 0.1$ and two H/D_j values, for example, $H/D_j = 4.0$ and 10.0, are selected.

Compared to natural convection only, the cocurrent jet removes heat from the tube at a rate twice as high as at the stagnation point ($\lambda = 0°$). Circumferentially after which, the local heat transfer decreases monotonically to the base point ($\lambda = 180°$). It should be noted that if the Archimedes number is substantially small, that is, $Ar_D << 1.0$, the round jet dominates the overall heat transfer behavior, making the natural convection insignificant, which is the case in most turbulent impinging jet studies.

The forced convective flow from the round jet interacts with the laminar boundary layer developed due to natural convection. The increased local heat transfer implies the thinning of the boundary layer. With the jet positioned at $H/D_j = 4.0$, a slightly higher heat transfer than that at $H/D_j = 10.0$ is achieved over the entire azimuth angle range whilst the distribution pattern remains unchanged.

In conventional impinging jet studies, the local heat transfer at a certain azimuth angle on the target cylinder is significantly elevated, forming a second peak due to transition from laminar to turbulent flow if positioned inside the potential core of the jet (Wang et al., 2014a,b). In the present study, the observed monotonic decrease in local heat transfer even at

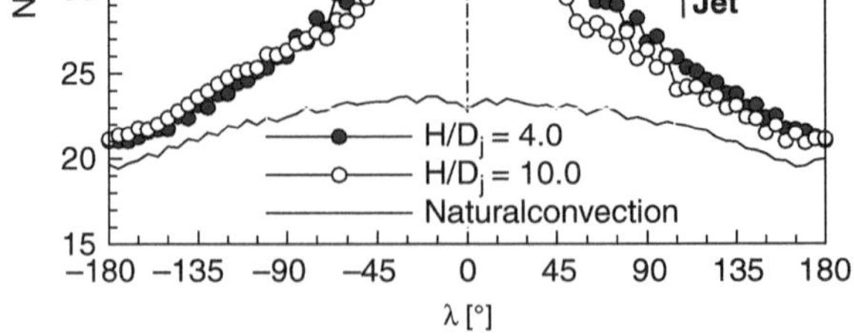

Figure 6.10 ***Local Nusselt number distribution at two selected*** **H/D_j** ***values for the cocurrent jet ($\alpha = 0°$).***

$H/D_j = 4.0$ (inside the undisturbed region) suggests that there is no transition to turbulent flow. It is thought that the jet is not strong enough to disturb the laminar boundary layer that is developed by natural convection and subsequently to trigger the transition. The slight decrease in the arrival jet velocity for $H/D_j = 10.0$ compared to the case for $H/D_j = 4.0$ results in a slight decrease in local heat transfer especially on the fore-cylinder region.

The area-averaged Nusselt number has been calculated to be 28.60 at $H/D_j = 4.0$ and 27.17 at $H/D_j = 10.0$. The former is 30% enhanced compared to the natural convection-only case whilst the latter is 23.4% enhanced.

C3. Countercurrent Jet-Buoyant Plume Interaction

Whilst natural convection removes heat from the tube, the jet is discharged at $\alpha = 180°$ according to the absolute coordinate system used in this study. However, for convenience, the local heat transfer distribution is plotted following the relative coordinate system denoting $\lambda = 0°$ for the stagnation point (as shown by the inset of Fig. 6.11). In this configuration, the direction of the jet is opposite to the buoyant plume from the tube—the countercurrent jet. The Archimedes number is also set at $Ar_D = 0.1$ and two H/D_j values, for example, $H/D_j = 4.0$ and 10.0 are selected.

With the countercurrent jet, the local heat transfer distribution shows a completely different pattern from that with the cocurrent jet since the jet needs to counteract the buoyant plume from the tube. At $H/D_j = 4.0$, the highest heat transfer exists at the stagnation point. After which, it decreases until $\lambda \sim \pm 60°$ and then slightly increases following the local heat transfer distribution of natural convection. At $H/D_j = 10.0$, even at the stagnation point, very marginal enhancement is obtained. At $\lambda \geq \pm 75°$, the local heat transfers of the two selected H/D_j values meet and follow the trend of

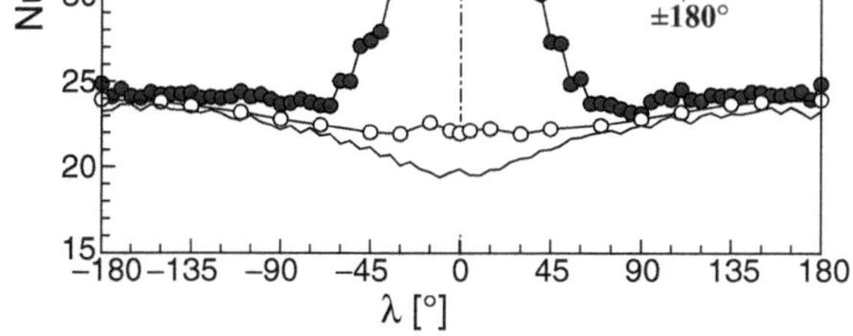

Figure 6.11 ***Local Nusselt number distribution at two selected*** **H/D_j** ***values for the countercurrent jet ($\alpha = 180°$).***

natural convection. This may indicate that the jet can disturb only the fore-tube region.

The area-averaged Nusselt number has been calculated to be 26.59 at $H/D_j = 4.0$ and 22.82 at $H/D_j = 10.0$. The former is 21% enhanced compared to the natural convection-only case whereas the latter is 3.7% enhanced, but this falls within the uncertainty range.

6.4 HEAT FLUX MEASUREMENT

6.4.1 Background

A heat flux sensor is an instrument used to measure the rate of heat energy transfer per unit area. It is a transducer which generates electrical signal (output voltage) which will be proportional to this measure of heat rate. The sensor is capable of measuring all three forms of heat transfer—conductive, convective, and also radiative heat transfer—on the surface or from the surface of a body.

In most heat transfer problems, both the temperature and the heat flux need to be quantified. When a given temperature drop ($\Delta T = T_2 - T_1$) is imposed across a solid material having a known thermal conductivity (k) and the two temperatures are measured by using, for example, thermocouples (Fig. 6.12a), the heat flux (q'') can be calculated according to Fourier's law, as:

$$q'' = -k\frac{\Delta T}{L} \tag{6.7}$$

where L is the thickness of the solid (plate) that the heat flow penetrates through. This is called the differential temperature heat flux measurement technique. Other available techniques can be found (Childs, 2001). It should be noted that if the solid thickness L is too small or the solid conductivity k is too large, the temperature difference ΔT becomes too small to be measurable by thermocouples (within their error bounds). Therefore, the above parameters should be properly selected to ensure a sufficiently large ΔT that lies outside the error bounds so that accurate heat flux values can be measured.

Fig. 6.12b shows a film-type heat flux gauge. It is composed of two foil-type thermocouples bound to either side of a thin film, for example, Kapton (Omega Engineering, inpress). With known thermal properties of the film [eg, thermal conductivity (k, W/mK)], the measurable temperature

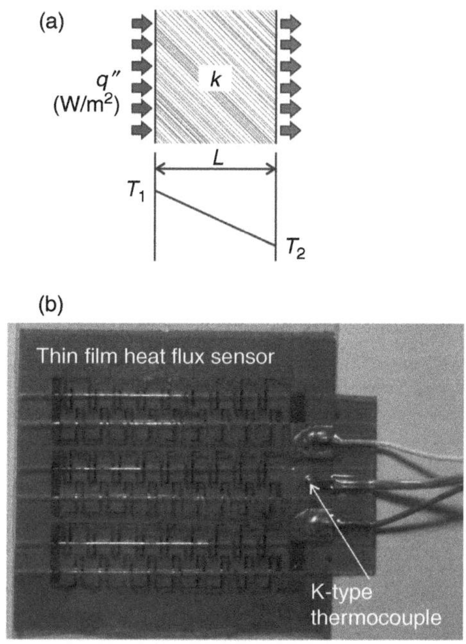

Figure 6.12 ***Heat flux gauge.*** (a) A schematic; (b) a foil heat flux sensor with K-type thermocouple (Omega™).

difference leads to the heat flux according to Fourier's law [Eq. (6.7)]. The total thickness of a film-type heat flux gauge is typically less than 200 μm.

In nature, it is very hard to achieve 100% efficiency in any experiment. There will always be a margin of error which must be accounted for. In many cases, measuring devices used cannot always be perfectly functional, hence the need for the experimentalist to calibrate the device. This improves the quality of one's results.

6.4.2 Calibration

Heat flux sensors record output voltage which is then used to determine heat flux. These heat flux sensors come with precalibrated data to easily convert the voltage readouts into heat fluxes. Each sensor will have different data based on its factory testing. The heat flux sensor reacts differently when operating outside of a normal room-temperature range. The sensor provided has been tested under different temperature ranges and then a multiplication factor graph is created and can be used to find a multiplication factor required for a certain temperature (Fig. 6.13). This plays a major role when the sensor is outside of the normal temperature range (0 to 100°C, or 32 to 212°F). This is also provided with the purchased heat flux sensor.

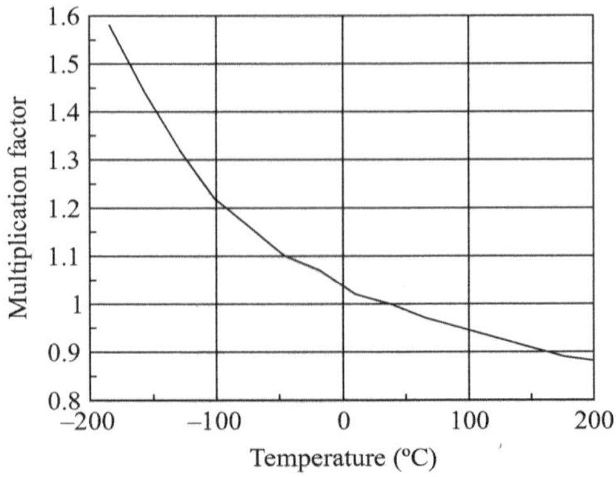

Figure 6.13 ***Heat flux gauge temperature correction factor (Omega™).***

In the case of achieving more accurate results and checking the functionality of the heat flux gauge, it is suggested that the heat flux sensor be calibrated. This can be done using Eq. (6.7), a block of known dimensions and known thermal conductivity, and the methods described previously.

6.4.3 Test Rig for Heat Flux Calibration

For accurate heat flux sensor calibration accounting for errors during in situ application, a test rig such as in Fig. 6.14a, b can be used. A sample block of specific 304 stainless steel with thermal conductivity of 16.2 W/mK is used to calibrate film-type heat flux sensors. The stainless steel block (dimensions of 100 mm × 100 mm × 50 mm) is placed inside a Perspex cubic container of low thermal conductivity (0.1 W/mK) for insulation. The wall thickness is set at 40 mm for maximum insulation. A thin copper plating is placed on top of the sample with thermal paste (Silicones™) and also at the bottom to ensure that the heat is transported uniformly through the sample block as it has a high thermal conductivity (~398.0 W/mK) and the paste helps reduce any thermal contact resistance. The conduction through this plate is assumed to be negligible due to its thinness ($T_2 = T_3$). The underside copper plate is in contact with the air in order to dissipate the transported heat through natural convection. The heating is introduced through a heating element attached to the top copper plate. The heating element is connected to a DC power supply which controls the voltage and current to impose constant heat flux. To minimize heat loss, a thick top covering made from Túfnol (0.25 W/mK) is used.

A T-type thermocouple is placed on the top of the steel block as well as the underside of the copper plating using a spray glue (3M™), to measure the temperatures T_1 and T_3 as shown in Fig 6.14a.

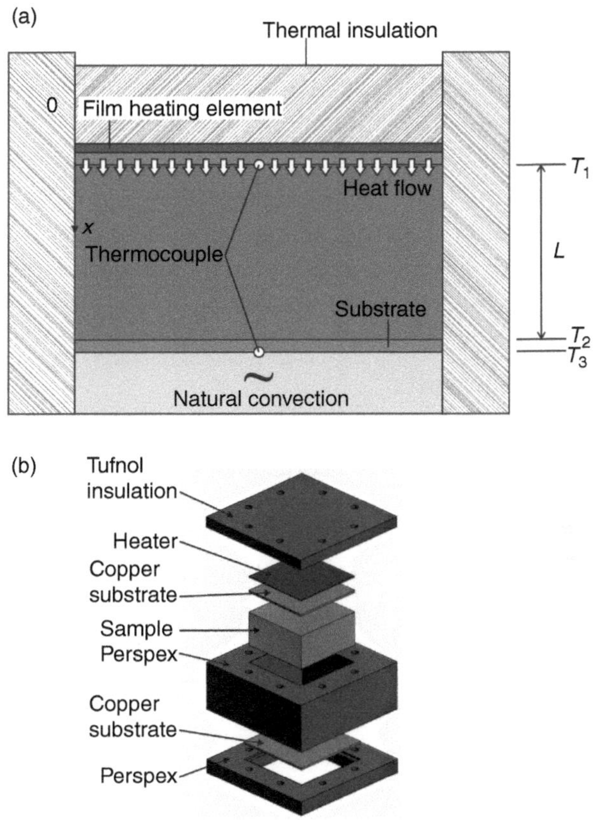

Figure 6.14 ***Schematic of a heat flux sensor calibration test rig.*** (a) Section view; (b) exploded view.

6.4.4 Calibration Method

To calibrate the heat flux sensors, selected input powers must be chosen to induce different heat fluxes. This is open to the user but must fall within the limitations of the heater used. Eight voltage values are selected to produce different input heat fluxes through the heater. The heat flux produced by the heater can be determined using electrical terms and the following equation:

$$q_h'' = \frac{V^2}{AR} \tag{6.8}$$

where V is the voltage supplied by the power supply in volts, A is the area of the heater in m^2, and R is the resistance of the heater in ohms.

The top and bottom T-type thermocouples are attached to a Fluke™ 52 II thermometer (Fig. 6.3a). The temperatures are recorded and then used together with the known thermal conductivity (16.2 W/mK) and Eq. (6.7) to calculate the heat flux passing through the sample block.

Two selected film-type heat flux sensors, Sensor 1 and Sensor 2, are connected to a DC voltage reader (Agilent™ in Fig 6.3b). The sensors are flush-mounted to the underside of the rig on the copper plate. To ensure the sensors were flush-mounted, a thin conducting tape was placed over the sensor to keep it attached firmly to the surface.

6.4.5 Calibration Results

To obtain effective calibration curves, the known heat flux (W/m^2) from Fourier's law [Eq. (6.7)] should be compared to the output voltage (μV) of the sensors. Fig. 6.15a, b show the results of the calibration from the current test rig for both sensors individually.

A linear relationship between the calculated heat flux and the output voltage from the sensors may be correlated. This is expected from the behavior of heat flux through a homogeneous solid and the expected response of the heat flux sensors. However, this is only the case for the tested temperature range of 0–100°C. As the heat flux increases, the voltage output measured from the sensor increases as a result. This is due to the expansion of the heat flux sensor wire inside the sensor, leading to a change in the potential difference.

The gradient from the linear relationship provides the calibration correction factor needed for that particular sensor. In the case of Sensor 1, Fig. 6.15a, the correction factor is calculated to be 7.96 W/m^2/μV with a resolution of 5%. For Sensor 2 (Fig. 6.15b), the correction factor is calculated to be 9.26 W/m^2/μV. Using the correction factors, measure the heat flux obtainable from an object by multiplying the recorded voltage output with the factor to obtain the heat flux from that object as:

$$q_{measured}'' = C_1(V)\,\mathrm{W/m^2} \tag{6.9}$$

where C_1 was correlated to be 7.96 W/m^2/μV for Sensor 1 and 9.26 W/m^2/μV for Sensor 2.

For example, if the recorded voltage obtained from Sensor 1 was 150 μV, then the following applies:

$$q_{measured}'' = 7.96 \times (150\,\mu V)\,\mathrm{W/m^2} \tag{6.10}$$

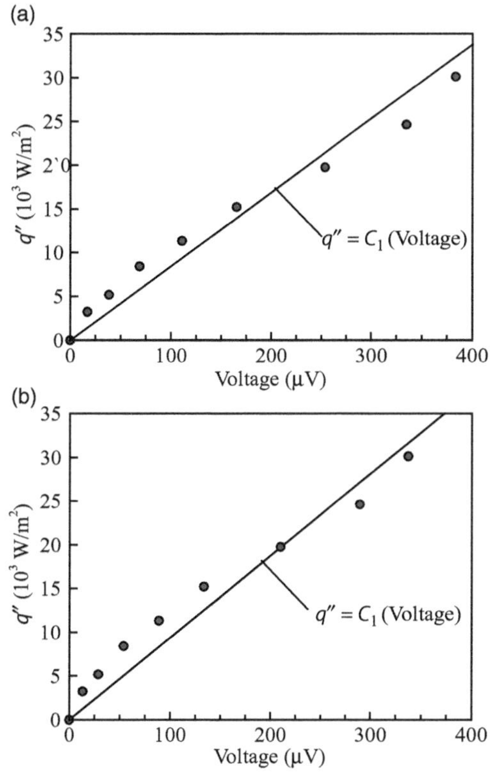

Figure 6.15 ***Calibration results for output voltage, V in microvolts versus resultant heat flux, q″ in W/m².*** (a) Sensor 1 having C_1 = 7.96 (W/m²)/(μV); (b) Sensor 2 having C_1 = 9.26 (W/m²)/(μV).

It should be noted that according to a manufacturer of these two film-type heat flux sensors, the provided correction factors are the same and close to 1.0 W/m²/μV. However, the present calibration results under a particular condition show different correction factors. Therefore, it is strongly advised that users need to calibrate them if possible.

REFERENCES

Atmane, M.A., Chan, V.S.S., Murray, D.B., 2003. Natural convection around a horizontal heated cylinder: the effects of vertical confinement. Int. J. Heat Mass Tran. 46, 3661–3672.

Beverly, T.L., 1995. About Temperature. Available from: <http://eo.ucar.edu/skymath/tmp2.html>

Childs, P.R.N., 2001. Practical Temperature Measurement. Butterworth-Heinemann, Oxford.

Coleman, H.W., Steele, W.G., 1999. Experimentation and Uncertainty Analysis for Engineers, second ed. John Wiley and Sons, New York, 120–124.

Dimmick, G.R., Chatoorgoon, V., Khartabil, H.F., Duffey, R.B., 2002. Natural-convection studies for advanced CANDU reactor concepts. Nucl. Eng. Des. 215, 27–38.
IAEA, 2000. Analysis of Severe Accidents in Pressurized Heavy Water Reactors. International Atomic Energy Agency Tecdoc 1594, 7–14.
Incropera, F.P., DeWitt, D.P., 1990. Fundamentals of Heat and Mass Transfer, third ed. John Wiley and Sons, Inc., pp. 530–552.
Kim, M., Yu, S., Kim, H., 2006. Analyses on fluid flow and heat transfer inside calandria vessel of CANDU-6 using CFD. Nucl. Eng. Des. 236, 1155–1164.
Maxim Integrated, 2007. Implementing Cold-Junction Compensation in Thermocouple Applications. Available from: <https://www.maximintegrated.com/en/app-notes/index.mvp/id/4026>
Misumi, T., Suzuki, K., Kitamura, K., 1999. Heat transfer and fluid flow of natural convection around large horizontal cylinders. Int. J. Heat Mass Tran. 42, 4093–4106.
Misumi, T., Suzuki, K., Kitamura, K., 2003. Fluid flow and heat transfer of natural convection around large horizontal cylinders: experiments with air. Heat Transfer – Asian Research 32, 293–305.
Morgan, V.T., 1975. The overall convective heat transfer from a smooth cylinder. Advances in Heat Transfer 111, 199–264.
Nandan, G., Sahoo, P.K., Kumar, R., Chatterjee, B., Mukhopadhyay, D., Lele, H.G., 2010. Experimental investigation of sagging and ballooning for a completely voided pressure tube of Indian PHWR under heat-up condition. World Acad. Sci. Eng. Technol. 60, 197–204.
National Instruments, 2011. Temperature Measurements with Thermocouples. How-to-Guide, Austin, TX.
Nguyena, T., Jaitly, R., Dinnie, K., Henry, R., Sinclair, D., Wilson, D.J., O'Neill, M., 2008. Development of severe accident management guidance (SAMG) for the Canadian CANDU-6 nuclear power plants. Nucl. Eng. Des. 238, 1093–1099.
Omega Engineering, Thermocouples. Available from: <www.omega.com>
Wang, P., Kahawita, R., Nguyen, T., 1990. Numerical computation of the natural convection flow about a horizontal cylinder using splines. Numer. Heat Transfer 17, 191–215.
Wang, X.L., Lee, J.H., Lu, T.J., Song, S.J., Kim, T., 2014a. A comparative study of single-/two-jet crossflow heat transfer on a circular cylinder. Int. J. Heat Mass Tran. 78, 588–598.
Wang, X.L., Yan, H.B., Lu, T.J., Song, S.J., Kim, T., 2014b. Heat transfer characteristics of an inclined impinging jet on a curved surface in crossflow. J. Heat Transf. 136 (8), , 081702-1-10.

CHAPTER 7

Surface Heat Transfer Mapping Using Thermochromic Liquid Crystal

T. Kim
School of Mechanical and Aeronautical Engineering, University of the Witwatersrand, Johannesburg, South Africa

7.1 INTRODUCTION

Temperature measuring devices such as thermocouples and RTDs only allow temperature at a point to be measured. To obtain a temperature field using these devices, they have to be spatially traversed. If one wants to calculate a temperature gradient on the traverse plane, the traverse grid should be "fine" enough to reduce errors associated with the temperature gradient. In such cases, a technique (device) that allows temperature field measurement would be preferable.

Thermochromic liquid crystal (TLC) is relatively cheap and easy to apply for the previous purpose. Unlike point temperature measurement techniques, the use of TLC can provide instantaneous temperature field measurement on a surface as well as convection heat transfer coefficient maps on the same surface through a transient method. Practically, the implementation of TLC measurement involves the application of TLC particles with a special adhesive on the surface where both the temperature and heat transfer coefficient are to be measured, followed by optically monitoring the responses of the TLC layer to temperature variations of the fluid flowing over the layer.

7.2 PROPERTIES OF TLC

TLC is a cholesteric liquid crystal that reacts to changes in temperature by reflecting different wavelengths of incident white light. With increasing temperature, the reflected light changes from "red" and passes through the visible spectrum to "violet" before reaching the optically inactive phase (liquid). The temperature when the TLC starts to display the red is called the "event" temperature whereas when it displays no visible wavelength after

Application of Thermo-Fluidic Measurement Techniques
http://dx.doi.org/10.1016/B978-0-12-809731-1.00007-1

the violet is called the "clearing" temperature. The changes of the reflected lights of the TLC are reversible unless they are exposed to excessively high temperatures for a prolonged time period.

The liquid crystal is isolated from the atmosphere by a microencapsulation process. This reduces the risk of contamination from its binder and the surface paints. More information on the physical properties of the liquid crystals and their application in heat transfer measurements can be found in the following references (Ireland et al., 1995, 1999; Camci et al., 1992; Kim, 2004; Kim et al., 2004; Gillespie et al., 1998; Wang et al., 1995, 1996, 1998; Hippensteele et al., 1983; Baughn et al., 1999).

7.3 APPLICATION AND CALIBRATION

7.3.1 Application and Illumination

To demonstrate how to use TLC for temperature measurements, the solution of the encapsulated liquid crystal (BM/R27C6W/C17-10, Hallcrest Inc.), premixed binder, and deionized water was sprayed on the target surface using an airbrush as shown schematically in Fig. 7.1. The pressure of the filtered compressed air was typically set to be 20 PSI. The distance from the nozzle of the airbrush to the target surface was approximately 150 mm although this value can be varied depending on the applied pressure in the airbrush. The typical thickness of a coated TLC layer is approximately 20 μm (Ireland et al., 1995). Subsequently, black backing paint (BB-G1, Hallcrest Inc.) was sprayed on top of the TLC layer with the same setup of the airbrush and the same distance to the surface. This setup is for the case that the TLC color changes are monitored through a transparent plate as shown in Fig. 7.1.

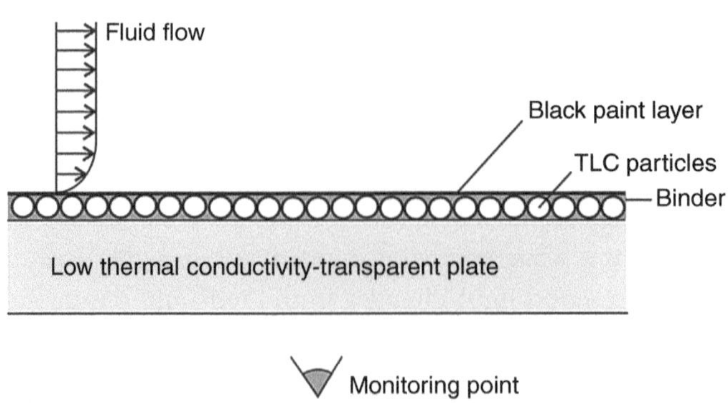

Figure 7.1 ***Schematic illustration of a TLC sprayed plate.***

At a given temperature, the reflected light may vary with the type, intensity, and incidence angle of the illumination. As reported in Camci et al. (1992), the dependence on the incident angle will not be significant if it is arranged to be within ±35° with respect to an axis normal to the surface. It is required that the source of illumination must not radiate additional heat to the surface. The illumination source used for the present measurements provides low thermal radiation with continuous cold white light (eg, Intralux dc-1100, Volpi Inc.).

7.3.2 Calibration

The calibration associated with a TLC measurement aims to determine the temperature from an instantaneous TLC image and the convection heat transfer coefficient from a series of TLC images that display temperature evolution in time. After the measurements, TLC images in the RGB (Red, Green, and Blue) space are converted into Hue, Saturation, and Intensity (HSI) space using a commercial program during post data processing. This is because in the RGB plane, there is no monotonic response (increase or decrease) of each element (Red, Green, and Blue) to the temperature variation. On the other hand, in the HIS plane, hue values increase linearly with temperature as illustrated in Fig. 7.2. Upon successful calibration, a one-to-one match between the hue values and the absolute temperature values is obtained.

To calibrate TLC in a similar configuration, it is sprayed on the inner surface of a water container that is made from 12 mm thick Perspex plates. A black backing paint is subsequently applied to enhance the reflectivity of the liquid crystal. Because the black paint is water soluble, the painted

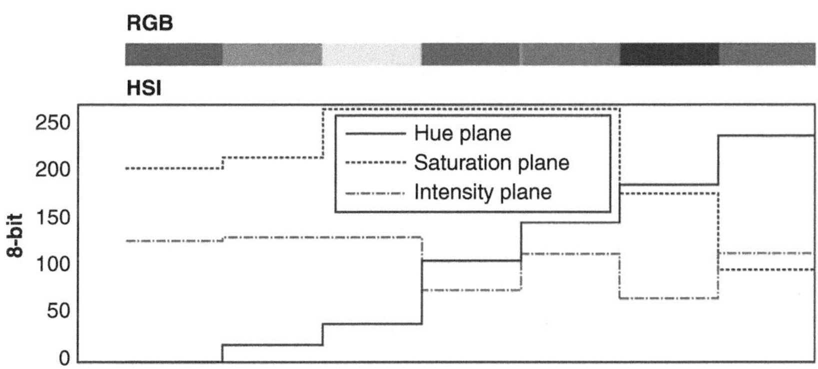

Figure 7.2 ***Illustration of conversion of image in Red, Green, and Blue (RGB) space into Saturation, Hue, and Intensity (SHI) space.***

surfaces are covered with a thin plastic film to prevent possible contamination of the layers. An electrical heater is used to raise the water temperature. A foil T-type thermocouple is placed between the Perspex plate and the TLC layer to monitor the surface temperature. Fig. 7.3a shows schematically the calibration setup (Kim, 2004).

The calibration results for the flat surface are presented in Fig. 7.3b. In most previous reported studies concerning the TLC technique, only a linear region from the whole of the calibration curve is used (Ireland et al., 1999; Gillespie et al., 1998; Wang et al., 1995, 1996, 1998; Hippensteele et al., 1983; Baughn et al., 1999). Although this limits the operating temperature range for a given TLC, the linear fitting appears to provide higher accuracy than other fitting schemes, for example, third-order polynomial curve fitting in the selected region. The calibration results in Fig. 7.3b are formulated with linear fitting as:

$$T_s = 0.057 \times \text{Hue} + 297.15\text{K} \tag{7.1}$$

where T_s is the calculated solid surface temperature and Hue is the converted hue value from the captured liquid crystal images ranging from 0 to 255 in the 8-bit system. The previous calibration is valid for $55 < \text{Hue} < 150$.

7.4 POST IMAGE ANALYSIS FOR HEAT TRANSFER COEFFICIENT

7.4.1 One-Dimensional (1D) Transient Heat Flow in a Semiinfinite Solid

Many transient heat transfer experiments assume that there is a sudden step change in thermal boundary condition imposed at the boundary surface of a semiinfinite solid. When the temperature of flow adjacent to the surface is suddenly raised from initial temperature, T_{init}, to current temperature, T_{ref}, transient 1D heat conduction occurs within the solid (which is originally also at a temperature, T_{init}). Under these circumstances, a 1D analytical solution is commonly used to find the convective heat transfer coefficient from the measured surface temperature history.

The time-dependent dimensionless temperature within the semiinfinite solid exposed to a sudden step change in fluid temperature may be expressed as:

$$\frac{T_s(x,t) - T_{init}}{T_{ref} - T_{init}} = erfc\left(\frac{x}{2\sqrt{\alpha t}}\right) - \left[\exp\left(\frac{hx}{k} + \frac{h^2 \alpha t}{k^2}\right) erfc\left(\frac{x}{2\sqrt{\alpha t}} + \frac{h\sqrt{\alpha t}}{k}\right)\right] \tag{7.2}$$

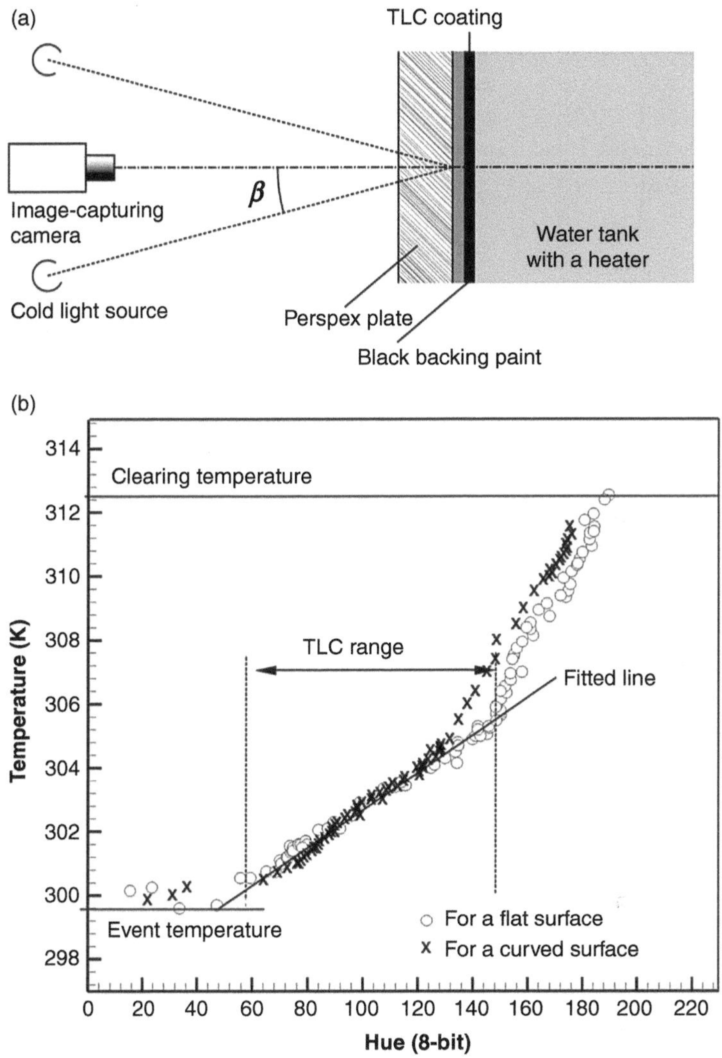

Figure 7.3 ***Calibration setup and results.*** (a) Schematic of the current calibration setup; (b) calibration results.

where $T_s(x,t)$ is the temperature within the semiinfinite solid, T_{init} is the initial solid temperature at $t = 0$, that is, $T_s(x,0) = T_{init}$, α is the thermal diffusivity ($= k/(\rho c_p)$), ρ is the density of the solid material, c_p is its specific heat in constant pressure, $erfc(\beta)$ is the commentary error function defined as $erfc(\beta) = 1 - erf(\beta)$, and $erf(\beta)$ is the Gaussian error function defined as:

$$erf(\beta) = (2/\sqrt{\pi})\int_0^{\beta} e^{-V^2}\, dV \tag{7.3}$$

Here, $\beta = h(\alpha t)^{1/2}/k$, h is the convection heat transfer coefficient to be determined, and V is a dummy variable.

Substitution of Eq. (7.3) into Eq. (7.2) yields the surface temperature $T_s(t)$ at $x = 0$ as:

$$T_s(t) - T_{init} = [1 - \exp(\beta^2)\cdot erf(\beta)](T_{ref} - T_{init}) \tag{7.4}$$

Eq. (7.4) can be explicitly solved for $T_s(t)$ if the values of α, h, k, T_{ref}, and T_{init} are given. On the other hand, if the history of the surface temperature is known, then the heat transfer coefficient h can be implicitly obtained from Eq. (7.4). This is the main concept employed in the present study.

7.4.2 Estimate of Adiabatic Reference Temperature

When the flow temperature is suddenly raised to the reference temperature T_{ref}, the evaluation of an adiabatic wall temperature T_{adw} (to be used as a reference temperature) is necessary. This is because the actual temperature of the flow may differ due to the effect of frictional heating. The adiabatic wall temperature is defined as:

$$T_{adw} = T_{ref} + R\frac{U^2}{2c_p} \tag{7.5}$$

where R is the recovery factor, U is the mainstream velocity, c_p is the thermal capacity of the fluid, and T_{ref} is the static temperature of the mainstream. If the flow velocity U is low, the adiabatic temperature then is nearly the same as the reference temperature, that is, $T_{adw} \sim T_{ref}$. In the following sections, the adiabatic wall temperature will be used as the reference temperature.

7.4.3 Estimate of Transient Test Time

Prior to the transient experiments, one must estimate the transient test time (t_{tran}) defined as the time taken for a single measurement. This is attributed to the use of the 1D heat conduction assumption in the semiinfinite solid plate of finite thickness (L), as shown in Fig. 7.4. That is, the single transient test time (t_{tran}) should be shorter than the time taken for the thermal pulse to penetrate the finite solid thickness (L). An analytical estimation for t_{tran} is

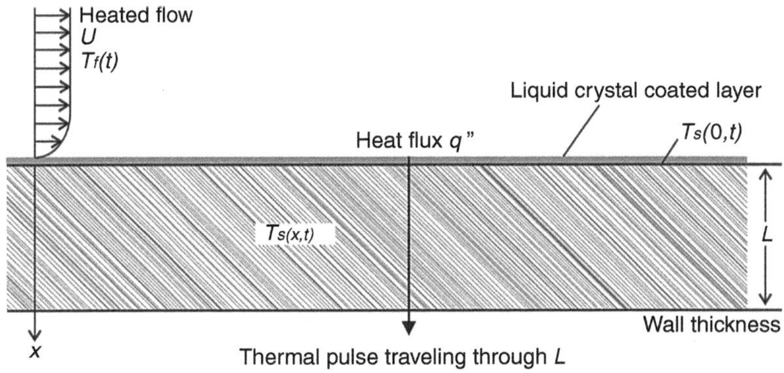

Figure 7.4 ***Schematic description of 1D semiinfinite plate equation.***

relatively straightforward. This can be performed by observing the temperature variation at the opposite side of the solid ($x = L$) in Eq. (7.2), given by:

$$T_s(L,t) = T_{init} + \left\{ erfc\left(\frac{L}{2\sqrt{\alpha t}}\right) - \left[\exp\left(\frac{hL}{k} + \frac{h^2 \alpha t}{k^2}\right) erfc\left(\frac{L}{2\sqrt{\alpha t}} + \frac{h\sqrt{\alpha t}}{k}\right)\right]\right\}\left(T_{adw} - T_{init}\right) \tag{7.6}$$

The thermal pulse traveling time can be calculated from Eq. (7.6) based on the properties of the chosen material. This has been found to be approximately 80 and 30 s for 10 and 6 mm thick Perspex plates across a wide range of the heat transfer coefficient, for example, h = 10 to 200 W/(m^{2}K).

As shown in Fig. 7.5, $T_s(L,t)$ remains the same as its original temperature (T_{init}) until the thermal pulse reaches the opposite side of the plate. Then, the temperature increases, depending on the magnitude of the thermal pulse (which is dictated by the heat transfer coefficient). The ratio of increase used in the current study is 1% variation from T_{init}, that is, $T_s(L,t) - T_{init} < 0.01\,T_{init}$, as used in Yan and Owen (2002).

In addition, a simple approximation of the transient time constant was reported by Ireland et al. (1995). This is expressed as $t_{tran} = L^2/(16\alpha)$ where α is the thermal diffusivity of the solid material. It has been established that this approximation correlates well with the measured transient time in the current study.

7.4.4 Modification of 1D Solution for Exponential Flow Temperature Rise

For the transient heat transfer experiments reported here, the air temperature actually increases exponentially with time, as:

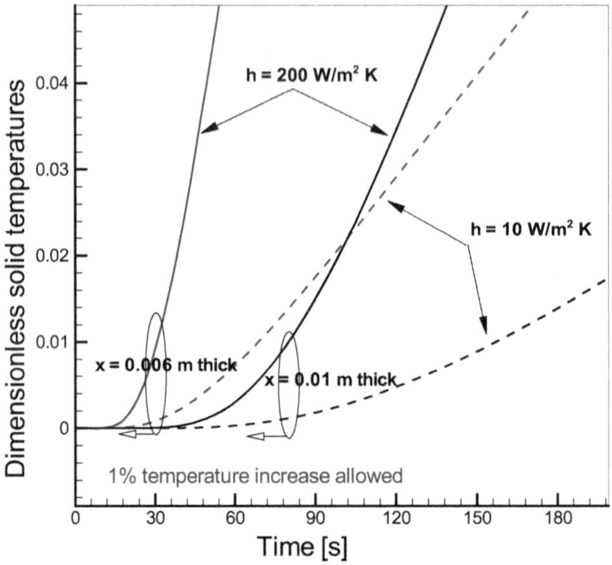

Figure 7.5 ***Estimate of transient running time where the dimensionless solid temperature is defined in Eq. (7.6).***

$$\Theta_f(t) = \frac{T_f(t) - T_{init}}{T_{adw} - T_{init}} = 1 - e^{-t/\tau_2} \tag{7.7}$$

where τ_2 is the time constant that depends on the ratio of the thermal capacity of the heating mesh to the heat transfer coefficient. As reported in Gillespie et al. (1998), for a given power input to the heating mesh, the time constant τ_2 decreases steeply as the flow velocity increases. To determine the surface heat transfer coefficient, it is necessary to measure the time constant for each experimental condition.

The exponential rise of the flow temperature, instead of the idealized step rise, means that the 1D solution presented in the previous section must be modified. According to Gillespie et al. (1998) and Gillespie (1996), the variation of temperature at the boundary surface of a semiinfinite solid when the flow temperature exhibits an exponential rise may be written as:

$$\begin{aligned} \frac{T_s(t) - T_{init}}{T_{adw} - T_{init}} &= 1 - \frac{1}{1+\beta_{\tau_2}^2} e^{\beta^2} erfc(\beta) \\ &- e^{-t/\tau_2} \frac{\beta_{\tau_2}^2}{1+\beta_{\tau_2}^2} \left\{ 1 + \frac{1}{\beta_{\tau_2}} \left[\frac{1}{\pi}\sqrt{\frac{t}{\tau_2}} + \frac{2}{\pi} \sum_{n=1}^{\infty} \frac{1}{n} e^{-n^2/4} \sinh\left(n\sqrt{\frac{t}{\tau_2}} \right) \right] \right\} \end{aligned} \tag{7.8}$$

where $\beta_{\tau 2} = h(\alpha\tau_2)^{1/2}/k_s$. If the measured time constant, τ_2, is short enough, say less than 1 s, the third term on the right hand side of Eq. (7.8) is negligibly small so that Eq. (7.8) becomes the original 1D solution, Eq. (7.2).

7.4.5 Determination of Surface Heat Transfer Coefficient

Fig. 7.6 illustrates schematically how the heat transfer coefficient is implicitly calculated by using the method discussed previously. Prior to the start of each experiment, the solid temperature, the fluid flow, and the TLC layer on the solid surface all have an initial temperature T_{init}. For all the cases considered in the present study, the value of T_{init} remains unchanged. When the heating mesh is activated, the flow temperature $T_f(t)$ increases exponentially as illustrated in Fig. 7.6. Then the TLC starts to display reflected light when the surface temperature reaches the event temperature, $T_{lc,event}$, of the TLC. In order to obtain the best results, it is advisable to start the measurements with $T_{lc,event} \leq T_{init}$. The temperature difference ($=T_{lc,event} - T_{init}$) varies depending on the initial (ie, ambient) temperature. If the temperature difference is sufficiently small, the color display of the TLC can be observed without a time lag (t_1).

The evolution of surface temperature measured by the TLC is presented in Fig. 7.6. During data processing, the 1D analytical solution is used to calculate the heat transfer coefficient. For comparison, analytical predictions of the surface temperature with either an exponential increase of flow temperature or a step rise of flow temperature are also included in Fig. 7.6.

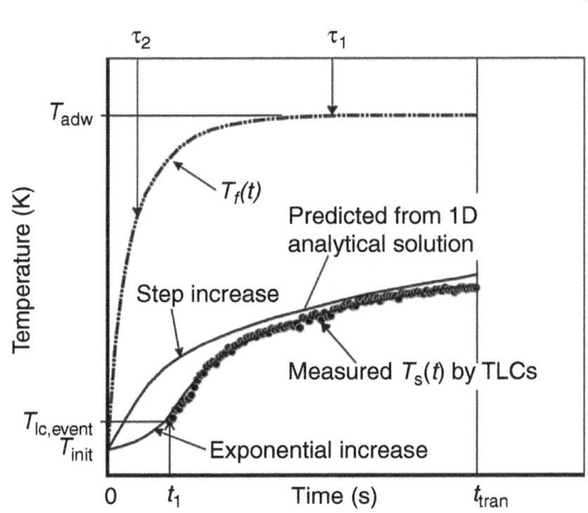

Figure 7.6 ***Illustration of temperature variations of fluid and solid surface in time.***

The effect of the time constant τ_2 on the overall trend of each curve for a prescribed heat transfer coefficient is illustrated in Fig. 7.7a, b.

A sample case for the reduction of a heat transfer coefficient by comparing measured temperature evolution with analytical predictions is shown in Fig. 7.8. The difference between the two fitted curves corresponding separately to idealized step rise of flow temperature and exponential increase of flow temperature is small, because the time constant τ_2 was set at 1.0 s. During the curve fitting process, a least squares error scheme was used to compare the difference between the temperatures measured by the TLCs and the predicted temperatures at each time step. When the value of the least squares error was within a specified error bound ($\varepsilon = 0.025\text{K}^2$), the process was stopped. The least squares error used during the matching process is:

$$E = \frac{1}{t_2 - t_1} \int_{t1}^{t2} \left(T_{measured}(t) - T_{eqn(3.6)}(t) \right)^2 dt < \varepsilon \qquad (7.9)$$

where t_1 and t_2 ($=t_{tran}$) are the time interval over which the curve fitting was carried out. Fig. 7.9 shows how to determine the heat transfer coefficient using the least squares error scheme.

This routine is repeated for all of the pixels in the images as illustrated in Fig. 7.10. Consequently, a complete heat transfer map is obtained. The total number of pixels selected for the present investigation was approximately 3.0×10^4, giving a spatial resolution of about 2.5 pixels mm^{-1} on the target surface.

Example 7.1: Endwall Heat Transfer Around an Inclined Cylinder

A. Background

An obstacle protruding from an endwall creates an adverse pressure gradient in the direction of the mainstream. As a result, the approaching boundary layer is separated from the endwall. This feature is called a "horseshoe vortex." In accordance with Kelvin's circulation theorem, the boundary layer vorticity cannot be destroyed. Instead, it is convected around each side of the obstacle to form the two legs of the horseshoe vortex. Numerous aspects associated with the horseshoe vortex have been studied due to its importance in engineering applications.

In some engineering applications, a single cylinder with both ends attached to endwalls is positioned at a certain angle of attack (inclination) to the mainstream flow (or the flow approaches the cylinder at an inclined angle). Fig. 7.11 shows high aspect ratio rotor blades which are connected

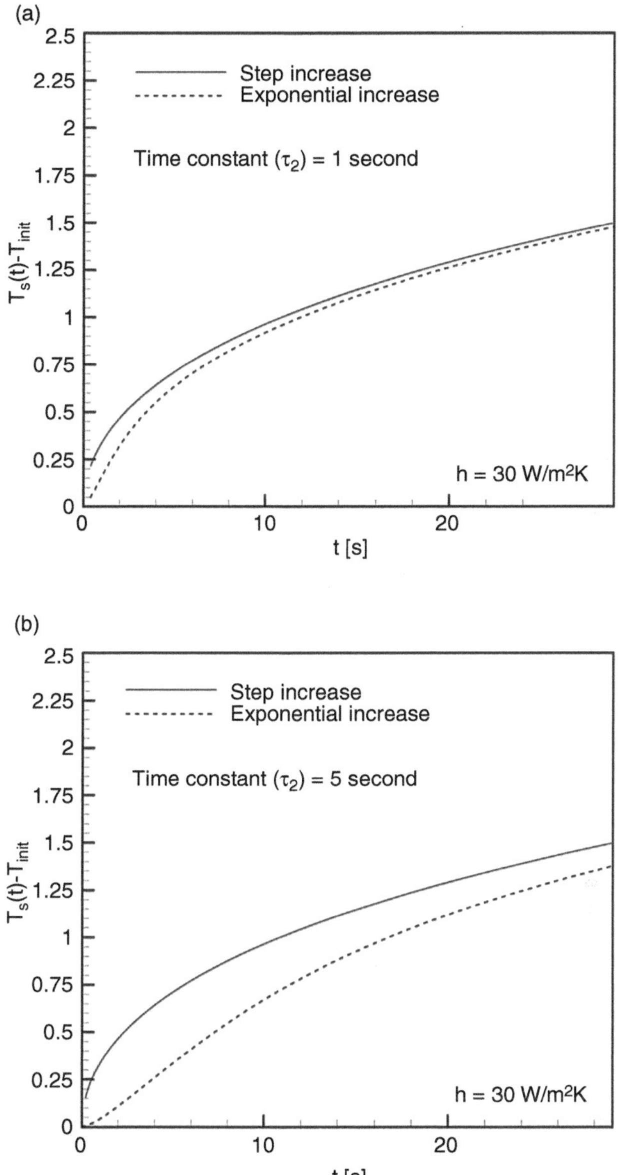

Figure 7.7 ***Comparisons of analytical predictions of temperature evolution with different flow temperature rise patterns (time constants).*** (a) For the time constant (τ_2) of 1 s; (b) for the time constant (τ_2) of 5 s.

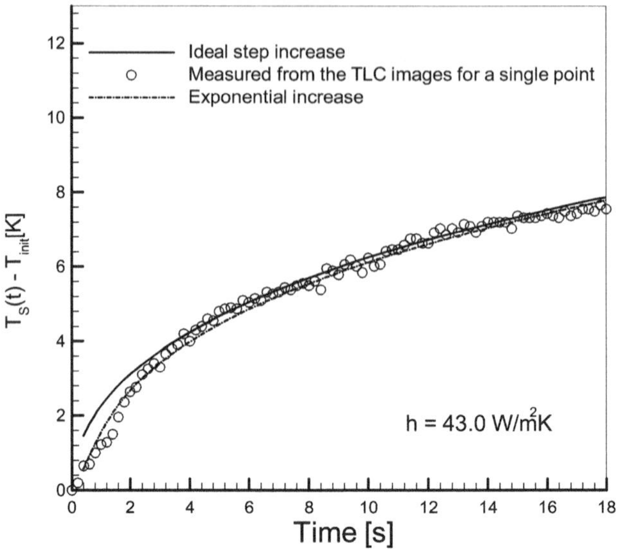

Figure 7.8 ***A sample case for the reduction of heat transfer coefficient by comparing measured temperature evolution with analytical predictions with the time constant (τ_2) set at 1.0 s.***

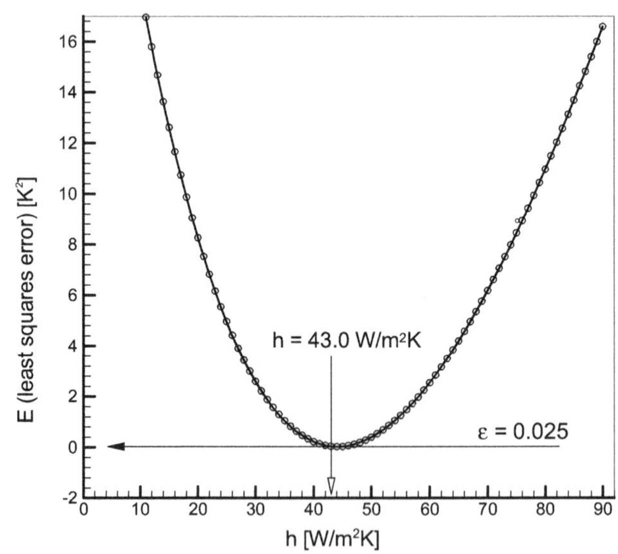

Figure 7.9 ***An example showing how to determine the heat transfer coefficient with least squares error where ε is the lower bound of the convergence.***

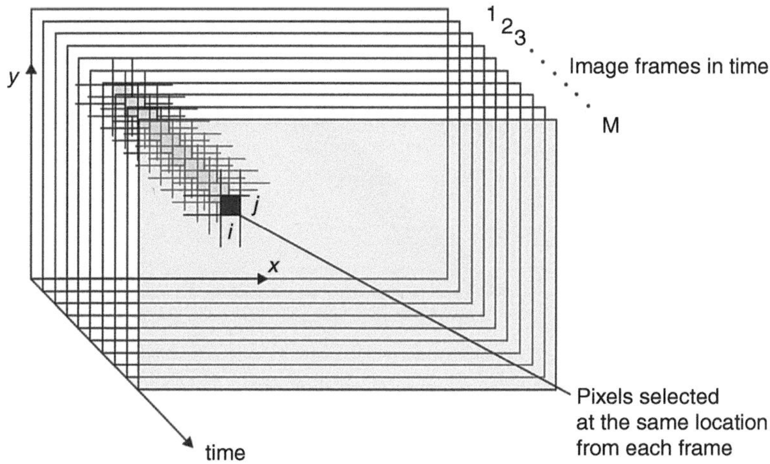

Figure 7.10 ***Illustration of image processing from captured RGB images in time.***

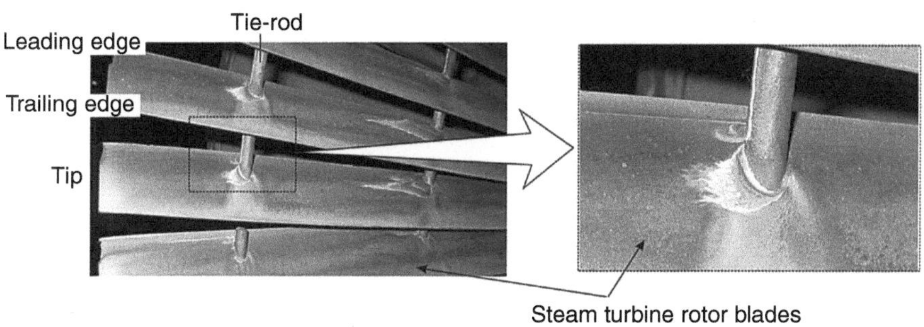

Figure 7.11 ***A photograph of steam turbine rotors showing rotor blades connected by a circular tie-rod.***

by circular tie-rods to reduce vibration during operation. In such a configuration, vortical flow patterns around the cylindrical rods become more complex due to flow features induced by the inclination. According to Choi et al. (2007), inclination-induced flow features around the single cylinder include (1) the skewed upstream main flow towards an endwall where lower pressure exists as it approaches the cylinder; (2) formation of a "jet-like flow" immediately downstream of the cylinder, followed by its impingement onto a region of an endwall plate downstream of the cylinder junction; and (3) a skewed separation line along the cylinder span from the cylinder axis. Among these, the skewed mainstream and jet-like flow may influence the horseshoe vortices formed at the junction.

This case study aims to articulate the variation of the size and strength (circulation) of the horseshoe vortices formed around a single cylinder subject to backward inclination and to examine the effect of such inclination

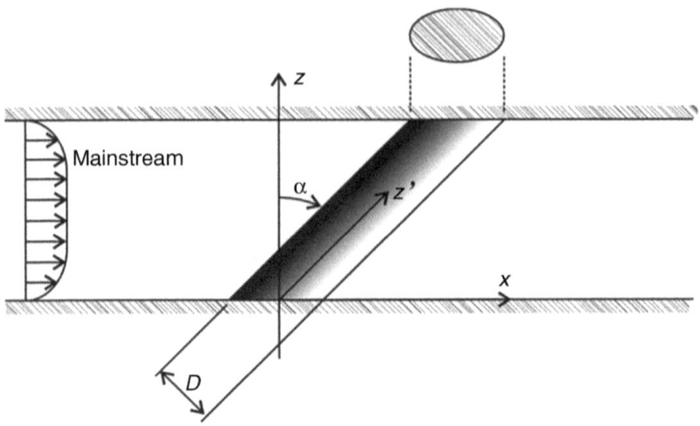

Figure 7.12 ***Configuration of a single cylinder with inclination.***

on endwall heat transfer. To this end, TLC is used to map the surface heat transfer coefficient around a cylinder whose angle to the mainstream varies from 0° (upright position) to 45° for a fixed Reynolds number of $Re_D = 1.0 \times 10^4$ (Fig. 7.12).

B. Test Setup

B1. Test Rig and Model

Flow past a single cylinder bounded between endwall plates is considered. Flow is supplied by a blow-down wind tunnel. Before the test section, a heating mesh which is capable of providing a sudden rise of flow temperature is placed. Prior to the heat transfer measurement, TLC is sprayed on the inner side of the lower endwall followed by a black paint layer.

The cylinder model is made of a low thermal conductivity tube ($k_s \sim 0.02$ W/mK) and has a length to diameter ratio of about 3, varied with its inclination. Details of the model are listed in Table 7.1.

Table 7.1 Parameters of Single-Cylinder Test Models for Flow Measurements in Wind Tunnel and Flow Visualization in Water Tunnel

Inclination angle α [degree]	Cylinder			Channel		
	Diameter D [mm]	Length l [mm]	l/D	Width W [mm]	Height H [mm]	Blockage ratio D/W
0	30	80.0	2.67	200	80	0.15
30		92.4	3.08			
45		113.1	3.77			

B2. Heating Mesh

During the data reduction process associated with the transient heat transfer experiments using the TLC technique, a 1D analytical solution assuming a semiinfinite solid is used. Consequently, it is necessary to employ a heating device that can provide the step input of the fluid flow temperature change. For this purpose the heating mesh is used as reported in Ireland et al. (1995) and Wang et al. (1995, 1998).

The heating mesh is made of stainless steel wire of 36 μm diameter. It is sandwiched between two Túfnol frames after being soldered to brass bus bars. The bus bars are used to connect the heating mesh to a low-voltage and high-current transformer. While air is passing through the heating mesh, the heat generated by the electrical current is transferred to the air. The heating mesh placed upstream of the cylinder model in the test section is shown in Fig. 7.13. The steady-state increase of air temperature ΔT caused by the heating mesh can be estimated using:

$$\Delta T = \frac{I^2 R}{\dot{m} c_p} \tag{7.10}$$

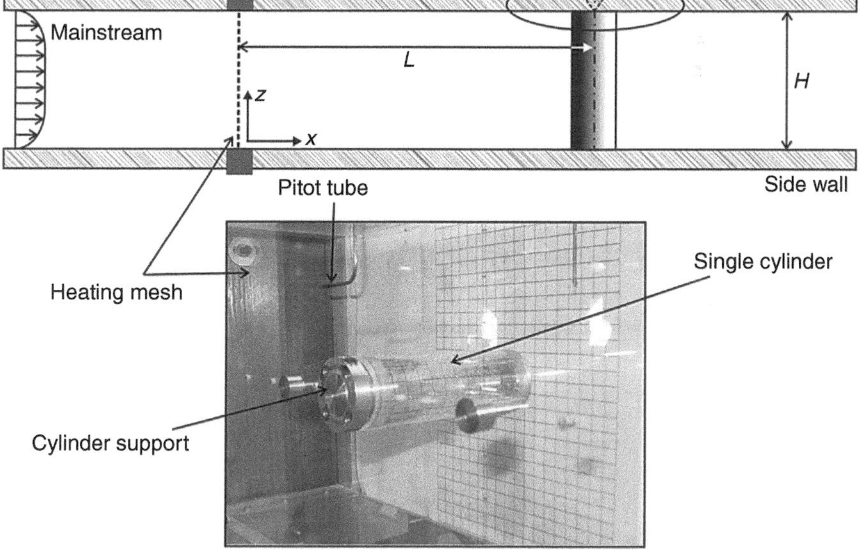

Figure 7.13 ***Schematic of the test rig for transient liquid crystal thermography.***

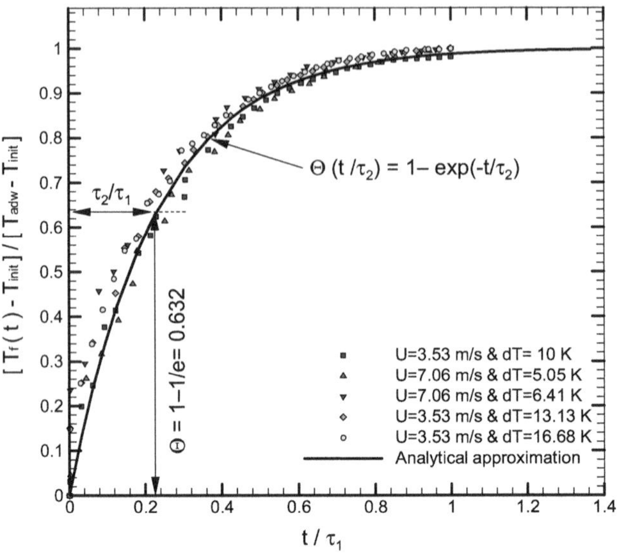

Figure 7.14 ***Thermal response of the heating mesh, varying with airflow velocity.***

where I, R, $\dot{m}$, and c_p are the imposed current and the electrical resistance of the heating mesh, the mass flow rate, and the thermal capacity of air, respectively.

During the transient measurements, the air temperature rises exponentially due to the thermal capacity of the mesh. The heating performance of the heating mesh used in this study is shown in Fig. 7.14. The fixed ΔT can be obtained by adjusting either the coolant mass flow rate or the input current to the heating mesh. In the current study, the mass flow rate was fixed at a desired flow velocity, and then the current was adjusted to give the required ΔT.

B3. Reduction of Surface Heat Transfer Coefficient

During the transient heat transfer experiments, the air temperature increases exponentially with time, described in nondimensional form by:

$$\Theta_f(t) = \frac{T_f(t) - T_{init}}{T_{ref} - T_{init}} = 1 - e^{-t/\tau_2} \tag{7.11}$$

where $T_f(t)$ is the inlet fluid temperature measured downstream of the heating mesh, T_{init} is the initial fluid temperature, T_{ref} is a reference temperature (to be defined later), and τ_2 is the representative time constant, or the time required for the dimensionless temperature to reach 0.632 (~ 3.0 s for Re_D = 1.0 × 10^4). Details associated with the heating mesh can also be found in Ireland et al. (1999).

When the flow temperature is suddenly raised to the reference temperature T_{ref}, evaluation of the adiabatic wall temperature T_{adw} is necessary (Yan and Owen, 2002), because the actual temperature of the flow may differ due to the effect of frictional heating. The adiabatic wall temperature is defined as:

$$T_{adw} = T_{ref} + RU^2/(2c_p) \tag{7.12}$$

where R is a recovery factor, U is the mainstream velocity, c_p is the specific thermal capacity of the fluid, and T_{ref} is the static temperature of the mainstream. Typical operating conditions in this study are: R (= $\mathrm{Pr}^{1/2}$) ~ 0.843 for air, U ~5.0 m/s, and c_p = 1003 J/kgK. The adiabatic temperature then becomes $T_{adw} \approx T_{ref}$ due to the low velocity of fluid flow and is typically assumed to be invariant in time (Cooper et al., 1975). In the following sections, the adiabatic wall temperature will be used as the reference temperature.

The nondimensional temperature variation at the boundary surface when the flow temperature exhibits an exponential rise is derived from Gillespie (1996):

$$\Theta_s(t,\tau) = \frac{T_s(t) - T_{init}}{T_{adw} - T_{init}} = f(\beta,\beta_\tau) \tag{7.13}$$

where

$$T_s(t) - T_{init} = f(\beta,\beta_{\tau_2})[T_{adw} - T_{init}]$$

Here,

$$f(\beta,\beta_\tau) = 1 - \frac{1}{1+\beta_\tau^2} e^{\beta^2} erfc(\beta)$$
$$-e^{-t/\tau} \frac{\beta_\tau^2}{1+\beta_\tau^2}\left\{1 + \frac{1}{\beta_\tau}\left[\frac{1}{\pi}\sqrt{\frac{t}{\tau}} + \frac{2}{\pi}\sum_{n=1}^{\infty}\frac{1}{n}e^{-n^2/4}\sinh\left(n\sqrt{\frac{t}{\tau}}\right)\right]\right\}$$

where $\beta = h\sqrt{\alpha t}/k$, $\beta_\tau = h\sqrt{\alpha\tau}/k$, h is the convection heat transfer coefficient to be obtained, α is the thermal diffusivity of the acryl plate, and τ is the time constant that depends on the performance of the heating mesh. For a prescribed value of the convection heat transfer coefficient h, Eq. (7.13) produces a temporal variation of the temperature difference $T(t) - T_{init}$.

C. Interpretation of Endwall Heat Transfer Distributions

The variation of the endwall surface heat transfer coefficient with inclination angles is displayed in Fig. 7.15. For heat transfer between endwall surfaces and mainstream flow, the cylinder acts to increase convection heat transfer like a turbulator or a protruded fin. Vortical structures such as the "horseshoe vortex" form around the cylinder. Fig. 7.15a shows the map of

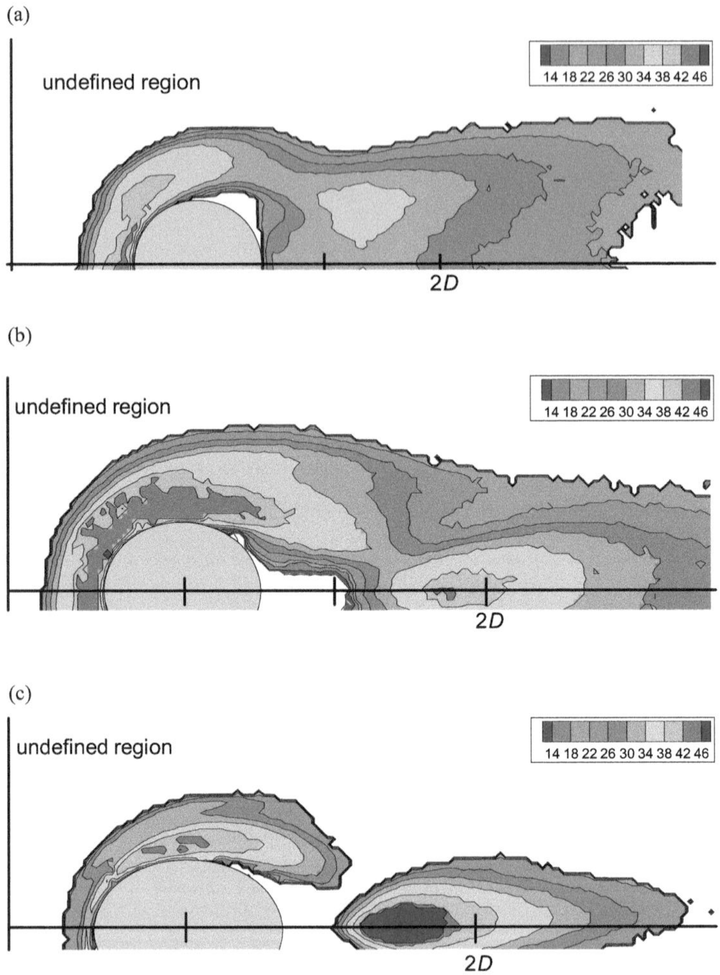

Figure 7.15 ***Distribution of endwall surface heat transfer coefficients (h) on the lower endwall at $Re_D = 1.0 \times 10^4$ (Choi et al., 2007).*** (a) $\alpha = 0°$ (upright position); (b) $\alpha = 30°$; (c) $\alpha = 45°$.

endwall convection heat transfer coefficient *h* for a cylinder with "zero" inclination angle. A high heat transfer region in front of the cylinder's leading edge results from the stagnation of flow. The white background area designated "undefined region" is caused by the nondetectable response of the TLC to low temperature change in time and magnitude in this region. Therefore, the convection heat transfer coefficient *h* could not be determined. Another white region near/on the cylinder indicates a separation bubble where a low-momentum fluid is trapped, leading to the lack of TLC response to the temperature change.

A relatively high heat transfer region (but not comparable to that of the stagnation region) behind the cylinder is observed at about 1.5D downstream from the cylinder axis and 0.5D away from the centerline. This region results from the interaction of the horseshoe vortex tails with the endwall surface.

The uncertainty, with a 95% confidence interval, associated with the heat transfer coefficient measurement is estimated to be less than 6.3%. The uncertainty has been estimated using the method of Coleman and Steele (1999). More details of the TLC technique can be found in Kim (2004), Ireland et al. (1999), Yan and Owen (2002), and Cooper et al. (1975). Prior to the heat transfer analysis, the buoyancy effect is estimated to be negligible due to the low operating temperature range, that is, less than 20 K in this study. A ratio of Gr_D to Re_D^2 is about 0.001 where Gr_D is the Grashof number defined as $Gr_D = g\beta\Delta TV/\nu^2$ (Yoon and Kim, 2002). Here, g is the gravity, β is the coefficient of thermal expansion, ΔT is the temperature difference, V is the representative volume (= cylinder diameter^3), and ν is the kinematic viscosity of air. In addition, since a low thermal conductivity material (Perspex, $k_s \sim 0.2$W/mK) was used for the cylinder, it is reasonable to assume that conduction through the cylinder is negligible in this study.

Fig. 7.15b, c display the endwall heat transfer coefficient contours of 30 and 45° inclined cylinder cases. The stagnation region of high heat transfer becomes thinner with increasing α. This may be associated with the change of vortical structures around the cylinder vertex including the "horseshoe vortex" as a result of the cross-sectional shape change (from circular to elliptical).

The most distinctive feature occurs at the aft cylinder. Upon inclining the cylinder, a peak value of the heat transfer coefficient behind the cylinder, for example, at 1.75D on top of the centerline, $y = 0$, becomes comparable to that at the upstream stagnation region for $\alpha = 30°$. A further inclination of the cylinder to $\alpha = 45°$ intensifies the magnitude of the heat transfer rate in that region which moves upstream to 1.5D (Fig. 7.15c). Moreover, the downstream high heat transfer region divides into two regions: one under the influence of the horseshoe vortex and the other affected by flow behind the cylinder. At the boundary between the two regions on the endwall surface, shear stress is so low that the TLC particles do not respond to the negligible temporal change of temperature there.

Example 7.2: Endwall Heat Transfer in a Sandwich Panel Cored by a Lattice-Frame Material

A. Background

A lattice-frame material (LFM) is a novel cellular material recently made possible by computer-based design and numerically controlled processing. The

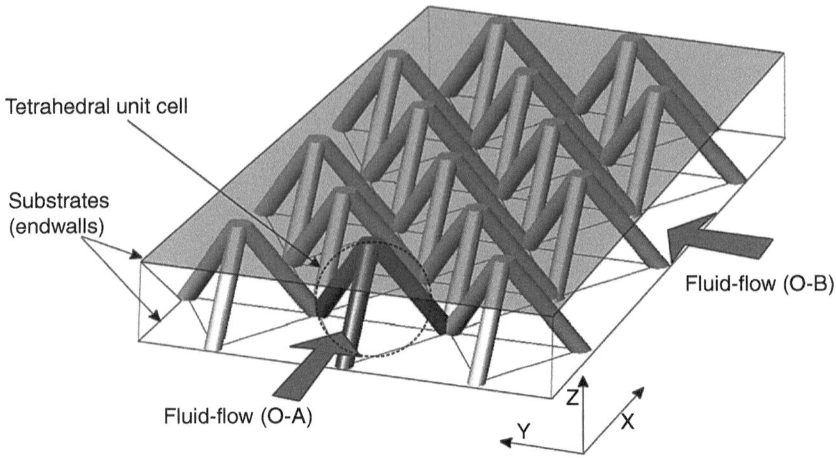

Figure 7.16 ***Single-layered lattice-frame material (LFM) model.***

LFM consists of a 3D network of cylindrical struts forming a tetrahedron as shown in Fig. 7.16. This tetrahedral unit cell is interconnected with other unit cells, which are periodically and spatially distributed to form a "periodic material." Recent developments in manufacturing techniques allow the diameter of these 3D lattice structures to vary from millimeters to tens of centimeters.

Unlike stochastic cellular materials (eg, metal foams), reported thermo-fluid research work on such a periodic material is very limited, because this material is relatively new. In order to obtain a desirable surface area density comparable to other conventional heat exchanger media such as fins and open cell metallic foams, relatively small cells of the order of a few millimeters are needed. New manufacturing technologies, for example, rapid prototyping, enable these scales to be achieved. The advantages associated with this periodic material over the fin arrays and open stochastic foams are: (1) mechanical response to external forces, for example, bending or stretching; and (2) precise control during the fabrication, minimizing imperfection of the structures, which leads to a decreased local thermal resistance.

This case study aims to show detailed endwall heat transfer distribution for two selected orientations – Orientation A (the most closed) and Orientation B (the most open) – of the aerodynamically anisotropic LFM, using TLC. The dimensions of the test sample are 0.127 m (width), 0.012 m (height), and 0.127 m (length). The diameter of the LFM strut is 2.0 mm. The ratio of strut length to diameter is 7.35 and the porosity of the test sample is 0.94.

B. Test Setup

The endwall surface heat transfer characteristics of the LFM are examined by employing the transient liquid crystal thermography. Fig. 7.17a shows the

(a)

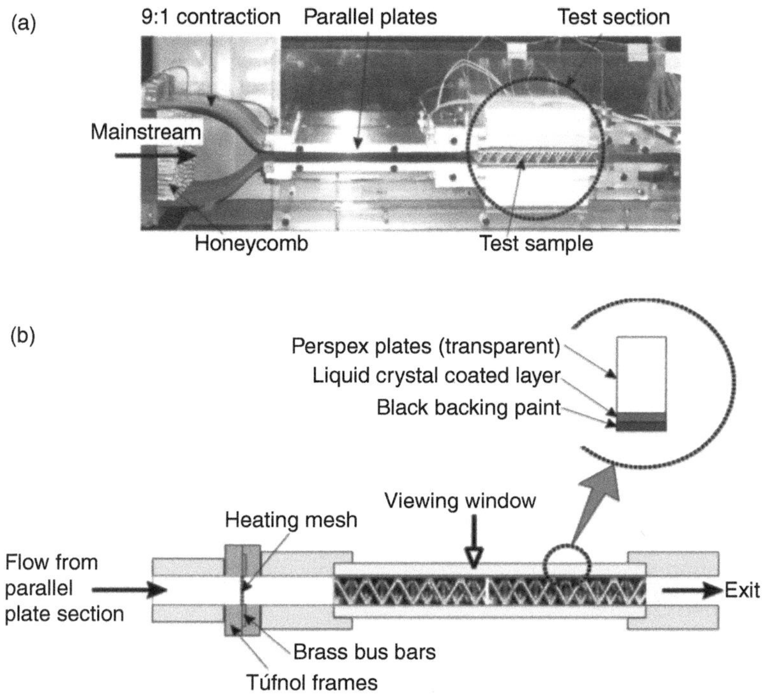

(b)

Figure 7.17 ***Test rig.*** (a) Photograph of small test rig for steady-state experiments; (b) schematic illustration of a modified version for transient experiments using TLC where a heating mesh was employed to provide sudden temperature increase of the coolant.

testing rig developed for the steady-state experiments, whereas Fig. 7.17b shows its modified version for the present transient tests. In general, the experimental apparatus consists of four main items: the coolant supply, the test section including the heating mesh at the inlet, the test sample, and the data acquisition system. Air at ambient conditions is drawn through the channel passage by a suction device. The coolant enters through a 9:1 contraction followed by a parallel section before it reaches the test section; the ratio of section length to channel height is 20.0.

C. Interpretation of Endwall Heat Transfer Distributions

C1. Orientation A

Fig. 7.18 presents the experimentally obtained Nusselt number distribution on the lower endwall where the Reynolds number based on unit cell length (d_p) is 4600. A short entry region is present, being no more than one unit cell in length. A similar endwall heat transfer pattern is repeated at each vertex beyond the first cell of each row. The endwall heat transfer patterns at each vertex are symmetric with respect to the center of the vertex and the flow direction. High heat transfer regions in front of and a short distance

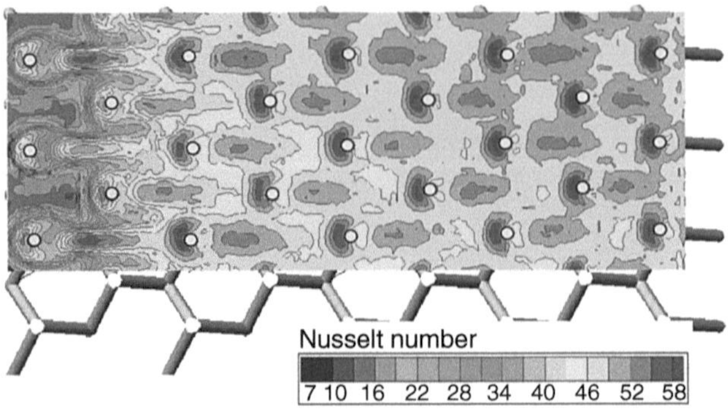

Figure 7.18 ***Contour of convection heat transfer coefficient on endwall surface for Orientation A (Kim et al., 2004).***

behind the vertices are observed. The former is due to the formation of a horseshoe-type vortex, whereas the latter results from the flow recirculation and flow reattachment of the wakes of the inclined and yawed struts.

C2. Orientation B

Fig. 7.19 presents the measured endwall Nusselt number distribution of Orientation B at a Reynolds number based on the unit cell length of 6000. The effect of the entry on the endwall surface heat transfer distribution appears to persist through most of the test model length although the effect diminishes rapidly after the first few cells. In general, two distinct endwall flow features are observed. One is due to the serpentine-shaped flow passage. The flow near the endwall surface associated with this flow pattern experiences less interference from the presence of the struts and vertices. This results in a lower heat transfer than elsewhere. The second feature is a

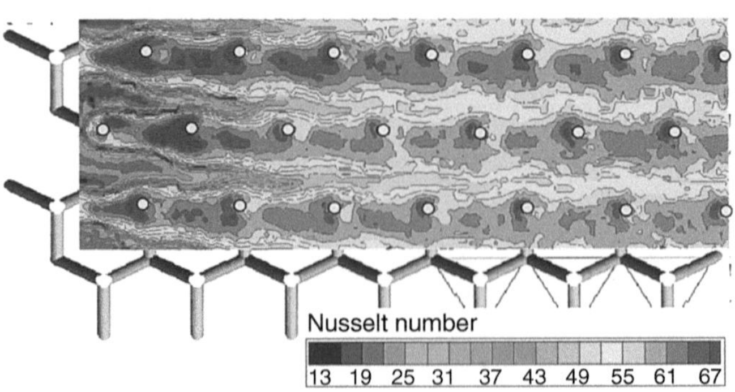

Figure 7.19 ***Contour of convection heat transfer coefficient on endwall surface for Orientation B (Kim et al., 2004).***

region of high heat transfer, which occurs near the vertices where there is a strong flow interaction.

Focusing on a single unit cell from a central location, three different types of struts form the unit cell in the orientation whilst two types of struts configure the unit cell in orientation A. The wake behind each vertex is skewed and asymmetric as a result of the asymmetric arrangement of the struts. In addition, the location of the center of the flow passage on the endwall surface appears to coincide with where a vertex is situated on another endwall plate. At this location, the least flow blockage is present. The high heat transfer occurs in front of and behind the vertex as in the case of Orientation A. However, in contrast to Orientation A, the horseshoe vortex is now skewed. The wakes from the connected struts cause the aforementioned high heat transfer region behind the vertex.

REFERENCES

Baughn, J.W., Anderson, M.R., Mayhew, J.E., Wolf, J.D., 1999. Hysteresis of thermochromic liquid crystal temperature measurement based on hue. J. Heat Transf. 121, 1067–1072.

Camci, C., Kim, K., Hippensteele, S.A., 1992. A new hue capturing technique for the quantitative interpretation of liquid crystal images used in convective heat transfer studies. J. Turbomach. 114, 765–775.

Choi, I.K., Kim, T., Song, S.J., Lu, T.J., 2007. Endwall heat transfer and fluid-flow around an inclined short cylinder. Int. J. Heat Mass Tran. 50 (5), 919–930.

Coleman, H.W., Steele, W.G., 1999. Experimentation and Uncertainty Analysis for Engineers, second ed. John Wiley & Sons, Inc., Hoboken, NJ.

Cooper, T.E., Field, R.J., Meyer, J.F., 1975. Liquid crystal thermography and its application to the study of convective heat transfer. J. Heat Transf. 97, 442–450.

Gillespie, D.R.H., 1996. Intricate Internal Cooling System for Gas Turbine Blading, D.Phil. Thesis, Department of Engineering Science. University of Oxford, Oxford, UK.

Gillespie, D.R.H., Wang, Z., Ireland, P.T., Kohler, S.T., 1998. Full surface local heat transfer coefficient measurements in a model of an integrally cast impingement cooling geometry. J. Heat Transf. 120, 92–99.

Hippensteele, S.A., Russell, L.M., Stepka, F.S., 1983. Evaluation of a method for heat transfer measurements and thermal visualization using a composite of a heater element and liquid crystals. J. Heat Transf. 105, 184–189.

Ireland, P.T., Wang, Z., Jones, T.V., 1995. Measurement Techniques: Liquid Crystal Heat Transfer Measurements, von Karman Institute for Fluid Dynamics, Lecture Series 1995-01.

Ireland, P.T., Neely, A.J., Gillespie, D.R.H., Robertson, A.J., 1999. Turbulent heat transfer measurement using liquid crystals. Int. J. Heat Fluid Fl. 20, 355–367.

Kim, T., 2004. Fluid-flow and heat transfer in a lattice-frame material, Ph.D. Thesis, Department of Engineering. University of Cambridge, Cambridge.

Kim, T., Hodson, H.P., Lu, T.J., 2004. Fluid-flow and endwall heat-transfer characteristics of an ultralight lattice-frame material. Int. J. Heat Mass Tran. 47, 1129–1140.

Wang, Z., Ireland, P.T., Jones, T.V., 1995. An advanced method of processing liquid crystal video signal from transient heat transfer experiments. J. Heat Transf. 117, 184–189.

Wang, Z., Ireland, P.T., Jones, T.V., Davenport, R., 1996. A color image processing system for transient liquid crystal heat transfer experiments. J. Turbomach. 118, 421–427.
Wang, Z., Ireland, P.T., Kohler, S.T., Chew, J.W., 1998. Heat transfer measurements to a gas turbine cooling passage with inclined ribs. J. Turbomach. 120, 63–69.
Yan, Y., Owen, J.M., 2002. Uncertainties in transient heat transfer measurements with liquid crystal. Int. J. Heat Fluid Fl. 23, 29–35.
Yoon, S.H., Kim, M.S., 2002. Investigation of circumferential variation of heat transfer coefficients during in-tube evaporation for R-22 and R-407C using liquid crystal. J. Heat Transf. 124, 845–853.

CHAPTER 8

Thermal Mapping Using Infrared Thermography

K. Kim*, S.S. Feng**
*School of Electrical and Information Engineering, University of the Witwatersrand, Johannesburg, South Africa
**School of Aerospace Engineering, Xi'an Jiaotong University, Xi'an, China

8.1 INTRODUCTION

Temperature is a quantitative measure of an object's degree of warmth resulting from the average kinetic energy of molecules of which the object is composed. The measurement of this physical quantity is important because it has many engineering applications and requires the use of thermal sensors (eg, thermostats, thermistors, resistance temperature detectors [RTDs], thermocouples) and temperature transducers, which are all intrusive. These sensors are frequently used in temperature measurement applications but cannot be used in situations where the objects that are required to be measured are small, inaccessible, and moving. Therefore, an alternative temperature measurement technique has been sought.

Infrared (IR) thermometry is one of the noninvasive temperature measurement techniques that permit temperature values both at a point and on a surface area to be obtained. IR thermometry systems (eg, IR camera) are commercially available, usually integrated with built-in software and a real-time display unit. Such a system is relatively expensive but user-friendly. In practice, the following questions may arise:

1. Why readers have to know the basic operating principles of the IR system; and
2. Why calibration is involved for a "presumably" ready-to-go system.

The answer to the former is obviously to acquaint the readers with how the IR system works. The latter is dependent upon how close your experimental setup is to the calibration conditions provided by the manufacturer. In other words, this procedure is essential to improve the accuracy of the IR system, as the conditions in your experimental setup will typically differ from those of the manufacturer.

Application of Thermo-Fluidic Measurement Techniques
http://dx.doi.org/10.1016/B978-0-12-809731-1.00008-3

The following sections provide a brief introduction to the theoretical background of how the IR system (eg, IR camera) works, and include:
1. Measuring temperature on metals and glasses
2. Energy detector
3. Advantages
4. A few parameters to be considered during calibration.

8.2 PRINCIPLES

IR is a type of electromagnetic radiation that lies between the visible end of the light spectrum and the microwave spectrum as shown in Fig. 8.1. It sits just after the red light of the visible light spectrum and has a wavelength which ranges from about 780 nm to 1 mm (Vollmer and Mollmann, 2010). Through experiments, it has been shown that all objects with a temperature above absolute zero emit radiation according to the temperature and properties of the object's surface. Hence, all radiation is made up of a continuous distribution of wavelengths from the electromagnetic spectrum, for example, objects at room temperature can only emit thermal radiation in the IR region and hence this radiation cannot be detected by the human eye (Gaussorgues, 1994).

The IR measurement technique uses this electromagnetic radiation to measure temperature. It provides a wide range of temperatures from approximately 50 to 6000 K. Radiation does not necessarily need a medium for transmission and can travel through a vacuum or a body of fluid. As shown in Fig. 8.1, the band between 0.1 and 100 μm is known as thermal radiation and that between 0.78 and 1000 μm as IR radiation.

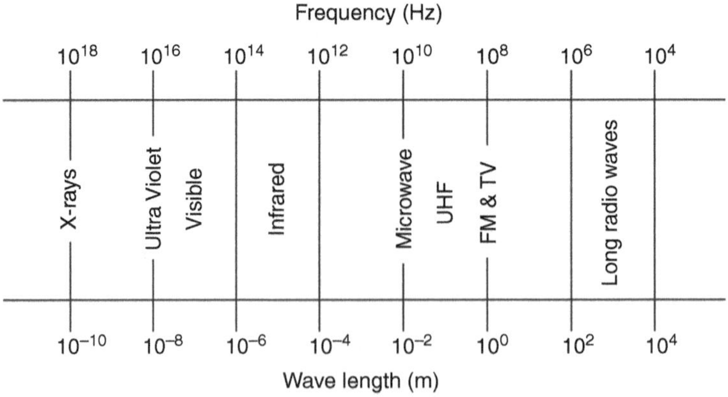

Figure 8.1 *The electromagnetic spectrum (Childs, 2001).*

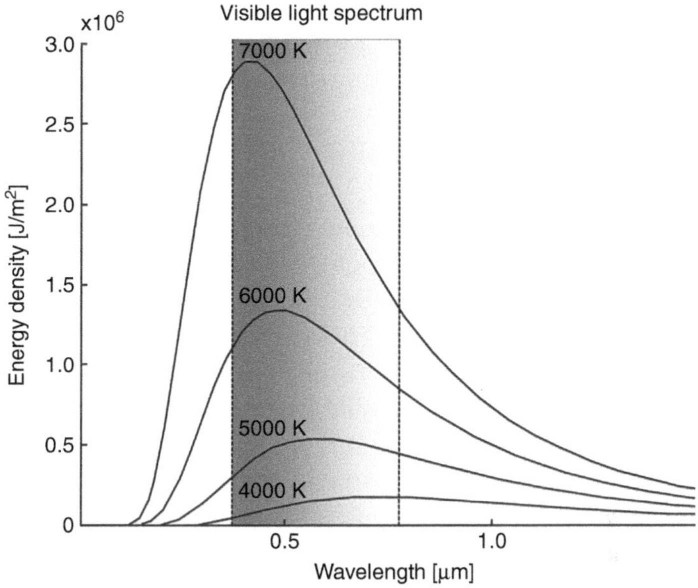

Figure 8.2 ***Blackbody radiation curve showing a relationship between wavelength and energy density (Jewett and Serway, 2010).***

A blackbody is defined as an ideal system or a theoretical object that absorbs all radiation incident on it and a body which emits the greatest amount of energy at a given temperature. It is a perfect absorber of radiation and any electromagnetic radiation emitted by a blackbody is called blackbody radiation. This is important as it enables temperature and wavelength to be calculated from the intensity and spectrum of blackbody radiation. Therefore, the following two laws for a blackbody: Wien's displacement law and the Stefan–Boltzmann law, were discovered due to the blackbody radiation assumption (Jewett and Serway, 2010).

Firstly, Wien's displacement law for a blackbody can be understood using a curve shown in Fig. 8.2. This curve shows that as the absolute temperature decreases, the peak wavelength of the blackbody radiation moves to a longer wavelength and lower energy density for a given object. An example of this is seen by objects that are at room temperature. The object does not appear to glow because the peak of the curve in Fig. 8.2 is not in the visible light spectrum but in the IR spectrum region. However, if the object becomes hotter, the peak of the radiation curve will move toward the visible light region. This can now be seen by the object eventually glowing red, then yellow, and then eventually blue (Jewett and Serway, 2010). This phenomenon can be explained by Wien's displacement law, which relates the maximum wavelength at which the curve peaks (λ_{max}) and the absolute temperature (T) of the object, given by:

$$\lambda_{max} T = 2.898 \times 10^{-3}\,\mathrm{mK}(\text{or } 2.898\text{ meter - Kelvin}) \tag{8.1}$$

This law enables the determination of the maximum wavelength of an object simply from the knowledge of the temperature or vice versa. This law is important in IR thermography as some IR cameras can only detect wavelengths within certain ranges. Therefore, Eq. (8.1) can be used to find the maximum and minimum temperature range of the IR camera to determine whether the IR camera is suitable for a specific temperature measurement application (Gaussorgues, 1994).

There is another important mathematical and theoretical expression linked to the radiation energy emitted from an ideal blackbody which has an absolute temperature of T, the so-called "Stefan–Boltzmann law," given by:

$$E_b = \sigma T^4 \tag{8.2}$$

where E_b is the total emissive power for a blackbody and σ is the Stefan–Boltzmann constant [= 5.67051 × 10^{-8} W/(m^2K^4)]. Eq. (8.2) states that the amount of radiation energy emitted in all directions over all wavelengths is directly proportional to the fourth order of the absolute temperature regardless of the prior knowledge of the energy-emitting blackbody's configurations (Childs, 2001). However, for a real surface, a constant (ε) called "emissivity" which is a partly material property and ranges from 0 to 1 (for a blackbody), deviating from a value for a given material, depending on many parameters including surface conditions, essentially needs to be introduced as:

$$E = \varepsilon \sigma T^4 \tag{8.3}$$

Emissivity is a measure of efficiency of an object that radiates thermal energy. It is defined as the ratio of energy emitted from an object's surface to that radiated from a blackbody at the same temperature. A blackbody is a perfect emitter of thermal energy with an emissivity value of one. The emissivity of an object, therefore affects how much energy an object emits and this information must be known for accurate temperature measurement; for example, if an object has an emissivity value of 0.77 (anodized aluminum) and its constant true temperature is 50°C, then the apparent temperature measured by the IR camera will be 38.5°C if the settings are not set correctly (Optotherm Thermal Imaging). Typical values of the emissivity for a variety of metallic and nonmetallic materials can be found in Childs (2001).

Eq. (8.3) is the working principle of an IR camera. An energy detector built in an IR camera measures the total emissive power emitted by an object and then inversely calculates the temperature using Eq. (8.3) based on the Stefan–Boltzmann constant and the object's emissivity. An IR camera cannot automatically determine the object's emissivity. A default value of 0.95 is typically used (eg, FLIR T-Series). The user is required to set the object's emissivity for accurate temperature measurements.

There is another parameter to be considered, which does not appear in Eq. (8.3), called "reflectivity." Reflectivity is the percentage at which electromagnetic radiation is reflected at a given wavelength by an object due to its physical properties. This mostly occurs from materials that are highly reflective such as polished aluminum, which reflects a large amount of radiation coming from the ambient. If the correct settings are not applied, the temperature measurements will deviate from the true temperature value (Optotherm Thermal Imaging, inpress). Some solutions to these problems, which mostly occur from using metals and plastics, will be explained in the following sections.

8.2.1 Measuring Temperature on Metals and Glasses

Measuring temperature on metals often requires the emissivity to be set on the IR camera as they tend to have a low emissivity because they reflect. This causes unreliable and inaccurate results, therefore it is important to select emissivity values of materials from tables, which are readily available. If there is a metal obstruction that is of no interest in the image, paint a thin layer of thermally black paint onto it so that it does not influence the measurement results (Gruner, 2003).

When measuring temperature on glasses, both reflectance and transmittance must be considered as thermal radiation can pass through. This allows the IR camera to capture the temperature on the surface and also at a depth. This is achieved by selecting different wavelength sensors for the IR camera (eg, use a 5.0 μm wavelength sensor for surface temperature measurements and a 1.0, 2.2, or 3.9 μm wavelength sensor for temperature measurements below the surface). Similarly to the case of the metals, if there is an obstruction due to the glass in the image, paint a thin layer of thermally black paint on it.

8.2.2 Energy Detector

The heart of the IR camera system that is required to detect the IR waves is the sensor (radiation detector). There are currently two types of radiation detectors that are commonly used to detect IR radiation: (1) thermal detectors, and (2) quantum detectors.

The thermal detector is commonly referred to as the uncooled microbolometer because this detector does not need cooling for the operation of the IR camera. The thermal detectors are constructed using metal compounds or semiconductor materials, and therefore cheaper to construct as these materials are commonly available. The thermal detectors also have a broader IR spectral response which allows various wavelengths in the IR region to be detected. However, thermal detectors are less sensitive and slower (milliseconds domain) than quantum detectors (microseconds domain) (Carlomagno and Cardone, 2010). A FLIR T-Series, which is used throughout this book, uses a thermal detector sensor.

Quantum detectors, another type of sensor, have a higher sensitivity and a lower response time than thermal detectors but require a strong cooling capability (eg, miniature Stirling-type coolers in cameras can achieve temperatures around 70 K). They are made from materials such as InSb, HgCdTe, and GaAs/AIGaAs whose sensitivity depends on photon absorption and, therefore are more expensive as these materials are not widely available. They also have lower IR spectral response which means they have a shorter range in which they can detect the IR radiation's wavelength (Carlomagno and Cardone, 2010). Quantum detector sensors are, therefore only used if a high sensitivity response is required at a specific wavelength. Hence, if high sensitivity and short response time are not required, the best IR camera choice would be the one with thermal detectors.

8.2.3 Advantages

IR cameras have been utilized successfully in industrial and research settings due to new innovations which have reduced costs, and increased the reliability of temperature accuracy. However, the following questions may still arise:

1. Why use an IR camera when there are cheaper alternatives such as thermocouples; and
2. What to keep in mind when using an IR thermometer.

The answers to the former question are (Gruner, 2003):

a. Time is saved because temperature measurement using an IR camera is faster than conventional thermocouples, with the speed at which measurement is taken in the millisecond range. Also, if a temperature map of a surface area is required, thermocouples can only take one point measurement at a time which is not efficient, whereas an IR camera can achieve a surface temperature map instantly without the need to take many measurements.
b. An IR camera allows for accurate temperature measurement of moving targets (eg, conveyor processes where contact measurement is impossible).
c. An IR camera can be used in situations where temperature measurements using contact measurement can be hazardous or inaccessible (eg, high-voltage equipment).
d. IR cameras can take measurements at high temperatures (in excess of 1300°C) where contact measurement is impossible.
e. Taking temperature measurements using an IR camera also allows for accurate temperature measurement as there is no interference where energy is lost from the target. This causes measurement errors when using thermocouples and needs to be considered as energy is lost through contact.

The answers to the latter question are (Gruner, 2003):

1. The target must be optically visible to the IR camera. A common mistake is the use of glass or Plexiglas/Perspex as an interface between the camera and the object that is to be measured (eg, a test rig made from an acrylic material such as Plexiglas/Perspex) as these materials are opaque to IR radiation. A glass made from quartz can be used as this is transparent to IR radiation.
2. It also must be noted that only the surface of an object can be measured using an IR camera. If an internal temperature measurement is required, it is advised to use a thermocouple, RTDs, and/or a thermostat.

8.3 CALIBRATION

8.3.1 Why in Situ Calibration?

Although a commercial IR system is supplied with built-in calibration data which were obtained under a specific industrial setup based on the "blackbody radiation" assumption, calibration close to the experimental setup of the users is still necessary before starting an experiment. Errors may occur, for example, systematic or random errors associated with different sets of test setups. The ideal setup is to calibrate the IR camera "in situ" (ie, in the same place as the setup where the camera is to be used), accounting for as many of the parameters as possible such as reflection from other objects other than the target point or surface, and viewing angle as a nonperpendicular viewing position may influence the temperature value recorded by the IR camera.

8.3.2 Calibration Setup

A commercially available model (Ti-50) from Fluke is employed for this section. Temperature values at a point on a surface flush-mounted foil thermocouple and those from the IR camera are compared whilst the viewing angle (relative angle between the target plate and the IR camera) is varied. The following two parameters are considered:

1. Systematic error
2. Viewing angle.

Fig. 8.3 schematically shows the test rig used for calibration. To create a uniform and constant temperature field on an aluminum (Al) block (even though the temperature at a point located at the center of the Al block is used), an etched foil heating element is attached to the side of

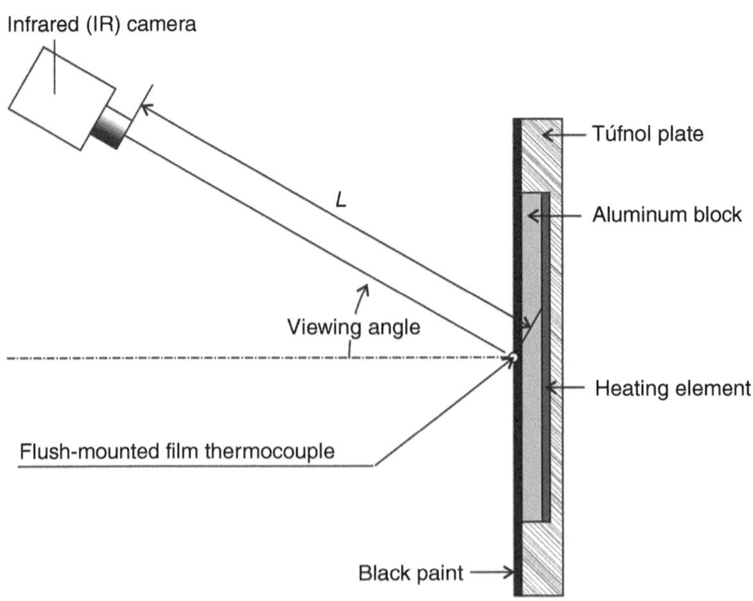

Figure 8.3 ***Schematic of calibration test rig.***

the block opposite to the side where the temperature field is monitored separately by a thermocouple and the IR camera. The IR camera focuses on the center of the Al block while the thermocouple is placed 5 mm off from the point on which the IR focuses. A black paint is sprayed on the Al block after a foil-type thermocouple (13 μm thick) is attached to the block surface.

The IR camera is positioned perpendicular to the Al block surface (ie, zero viewing angle) at a distance of $L = 0.385$ m (Fig. 8.3). Prior to calibration, the uniformity of temperature on the Al block along the horizontal axis is confirmed, with a deviation less than ±0.05°C that is within the error bound specified by the IR camera manufacturer. The whole calibration setup is placed in a black tent so that any possible radiation from the surroundings may be minimized. To calibrate the viewing angle, the relative angle of the Al block to the IR camera (the position of which remains unchanged) is varied horizontally to exclude the buoyancy effect.

8.3.3 Systematic Error (Comparison With a Precalibrated Thermocouple)

Fig. 8.4 presents the deviation of temperature measured by the IR camera from that obtained from the precalibrated foil thermocouple, with the viewing angle and the focal length fixed at 0° and 0.385 m, respectively. It is seen that the temperature measured by the IR camera is underestimated, increasing in a linear manner but systematically deviating from that obtained from the thermocouple as:

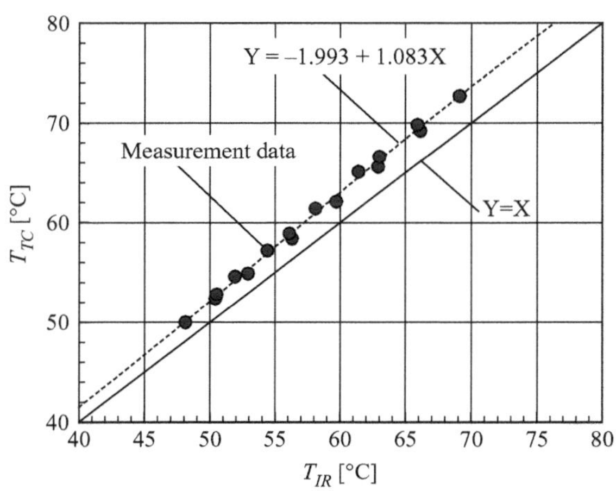

Figure 8.4 ***Measured systematic errors.***

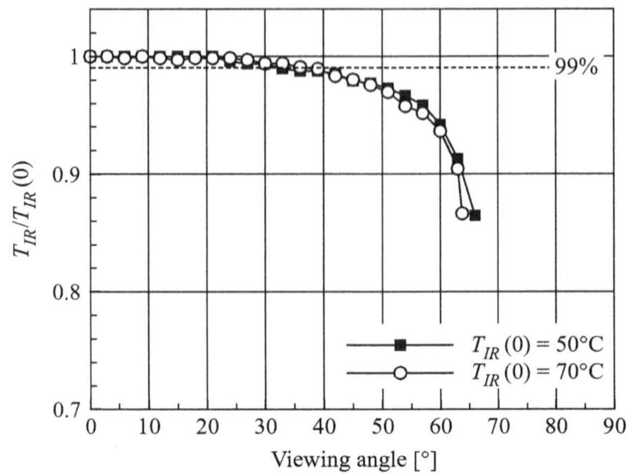

Figure 8.5 ***Variation of temperature readout from IR camera with viewing angle.***

$$T_{corrected} = (1.083T_{infrared}) - (1.993) \tag{8.4}$$

8.3.4 Effect of Viewing Angle (α)

Whilst the focal length is fixed at $L = 0.385$ m, the viewing angle α is systematically varied with an increment of 2.5° up to 65°; the normal position is chosen to be coincident with $\alpha = 0°$. Two different temperature setups on the Al block, for example, 50 and 70°C, are tested. The results are presented in Fig. 8.5.

For relatively small viewing angles ($0° \leq \alpha \leq 40°$), the temperature measured by the thermocouple is identical to that obtained with the IR camera, with the temperature at $\alpha = 40°$ only about 1.0% lower than that at $\alpha = 0°$. On the other hand, a further change in the viewing angle up to 65° leads to a significant deviation as shown in Fig. 8.5.

8.4 ESTIMATING HEAT FLUXES

IR thermography provides a surface temperature map remotely. Such a temperature map (or distribution) obtained is, in fact, as a result of heat that is transferred from/through an object of interest. For example, the temperature map on a fire protection panel's outer surface (ie, $z = L$) in Fig. 8.6 results from the heat flux that radiates from the fire. In addition to the temperature map on the outer panel surface, if the distribution of the heat

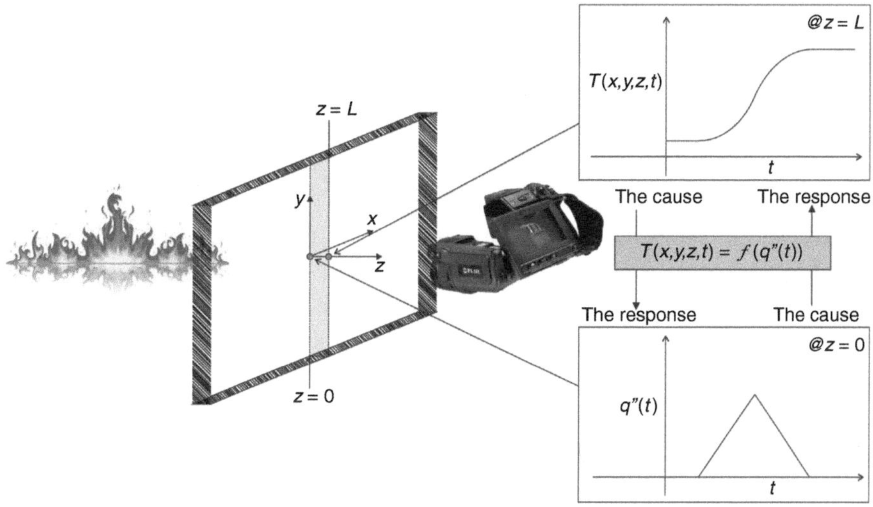

Figure 8.6 ***Schematic on an inverse method for estimating heat flux based on the temporal variation of surface temperature at a point.***

flux on the panel surfaces which can be either uniform or nonuniform is of interest, heat flux sensors need to be used. The imposed heat flux on the inner panel surface is conducted through the panel thickness, which is seen as the temperature map by the IR camera.

A question arises as to whether the temperature map obtained by IR thermography can be used for this purpose. Thus, numerous studies have been devoted to developing methods to estimate such heat flux at a point or its distribution (or map) on the outer surface, based on certain characteristics of the temperature(s) measured on the outer surface—inverse problems. In general, solving inverse problems requires exhaustive mathematical derivations with assumptions and, often, linearization and a series of numerical simulations.

Numerous schemes have been proposed to solve such inverse problems including:

1. The specified sequential function method based on the least squares method and Duhamel theorem (Beck et al., 1985), which is one of the schemes most used,
2. An integral method, adopting time steps by Stolz (1960),
3. A finite difference method applying the least square method by Frank (1963), and
4. A hybrid method by Cheng and Chang (1990).

For other methods being used to solve inverse problems in 1D heat conduction, the thorough review by Kumar et al. (2012) may be useful.

To exemplify the utilization of these schemes for the inverse problem to estimate heat flux based on the measured certain characteristics of temperature(s), one method proposed by Fernandes et al. (2015), employing the analogy of dynamic systems, is briefly discussed. The focus of this method is to derive the transfer function (or the impulse response) for the 1D heat conduction problem as a dynamic system by using the standard X22 model (Cole et al., 2010). Firstly, the direct method is used to solve the unknown surface temperature using Green's function and the standard X22 model, based on the prescribed heat flux values. The integral solution of Green's function is provided (Cole et al., 2010) as:

$$T(z,t) = T_0 + \alpha \int_0^t G_{x22}(z,t\,|\,0,\tau)\frac{q''(\tau)}{k}\,d\tau \tag{8.5}$$

where T_0 is the initial temperature, which is typically an adiabatic temperature, α is the thermal diffusivity, k is the thermal conductivity, t is the time variable, q'' is the prescribed heat flux, τ is a dummy time variable used for the integration process, and G_{x22} represents Green's function. Here, the subscript $_x$ indicates the axis of 1D conduction.

Through mathematical manipulation, the temperature distribution for the 1D conduction problem may be expressed as:

$$T(z,t) = T_0 + \frac{\alpha}{kL}\int_0^t q''(\tau)\,d\tau + \frac{2\alpha}{kL}\sum_i^M e^{-\left(\frac{i\pi}{L}\right)^2 \alpha t}\cos\left(\frac{i\pi z}{L}\right)\int_0^t e^{\left(\frac{i\pi}{L}\right)^2 \alpha t} q''(\tau)\,d\tau \tag{8.6}$$

where L is the thickness along the z-axis (Fig. 8.6), $i = 1,2,3\ldots M$, and M is the total number of convergence steps based on the specified convergence criteria, ε.

In a real situation, the heat flux is essentially unknown but the temperature distribution is known, for example, measured by an IR camera. Therefore, the heat flux needs to be calculated inversely based on the temperature distribution – the inverse method. Using the convolution integral (Fernandes et al., 2015) and the dynamics theory, the temperature output based on the prescribed heat flux input is given as:

$$T(z,t) = I(z,t) \star q''(t) = \int I(z,t-\tau)q''(\tau)\,d\tau \tag{8.7}$$

where $I(x,t)$ represents an impulse function for the system. For a "discrete" heat flux, the heat can be defined in terms of the Dirac delta function as:

$$q''(t) = \delta(t) \tag{8.8}$$

This eliminates the need to evaluate the integral, as the Heaviside function can be applied, resulting in:

$$T(z,t) = I(z,t) \tag{8.9}$$

Now relating Eqs. (8.5) and (8.9), the impulse response may be written as:

$$I(z,t) = T(z,t) - T_0 = \alpha \int_0^t G_{z22}(z,t \mid 0,t-\tau)\frac{q''(\tau)}{k}d\tau \tag{8.10}$$

Comparing Eq. (8.9) with Eq. (8.10) and replacing $q''(\tau)$ with $\delta(\tau)$, the impulse response for the specified 1D heat conduction problem becomes independent of the heat flux, yielding:

$$I(z,t) = \frac{\alpha}{k}G_{z22}(z,t \mid 0,t) = \frac{\alpha}{kL} + \frac{2\alpha}{kL}\sum_i^M e^{-\left(\frac{i\pi}{L}\right)^2 \alpha t}\cos\left(\frac{i\pi z}{L}\right) \tag{8.11}$$

Finally, once the general analytical model of the impulse response [Eq. (8.11)] has been derived, the solution of the inverse problem through further mathematical implications such as the dynamic systems theory and the Laplace transformation may be established as follows:

$$q''(t) = \frac{1}{I(z,t)} \times T(z,t) \tag{8.12}$$

For the actual implementation of this method, an input discrete heat flux (for mathematical simplicity) and arbitrary thermal physical properties need to be prescribed. The derived general temperature distribution in Eq. (8.6) becomes a particular distribution based on the prescribed heat flux input with the substitution of thermophysical parameters. To validate the developed inverse method, the simulated temperature distribution is substituted into Eq. (8.12) as well as Eq. (8.11). The generated heat flux has been found to be in good agreement with their prescribed heat flux.

In summary, with the measured temperature variation and known thermophysical properties of the solid, the heat flux can be inversely estimated.

Example 8.1: Thermal Boundary Layer Development Along a Porous Channel

A. Description of Problem

Boundary layers are developed along the channel walls when a fluid is forced to flow inside a rectangular channel. The velocity profile normal to the walls is determined based on the cross-sectional shape of the channel, the stream-wise location in the channel, and the flow conditions (eg, developing versus fully developed regimes). These features could be qualitatively visualized with relatively simple techniques, for example, hydrogen bubble generation (Smits and Lim, 2003), and quantified by traversing a Pitot tube normal to the channel wall at a selected stream-wise position.

On the other hand, a thermal boundary layer which is governed by the momentum boundary layer mentioned above is not so easy to visualize. Amongst the few available techniques, traversing a probe with a built-in thermocouple normal to the channel wall (which is, for example, heated) enables us to obtain a temperature profile with which the thermal boundary layer may be determined.

This example presents a simple way to visualize, in a qualitative manner, the thermal boundary layer developed from a heated rectangular channel wall. To this end, fluid flow across a rectangular channel filled with FeCrAlY foam having open cells is considered. Heat conducted from the channel is spread through the foam ligaments, which is then removed by the forced convective flow. It is assumed that the foam ligaments and fluid adjacent to the ligaments are in thermal equilibrium, implying that the temperature of foam ligaments to be monitored is the same as that of the fluid adjacent to the ligaments. Of concern is the thermal boundary layer developed in the rectangular channel where the fluid flow is highly disturbed by the filled metal foam. It should be noted that the thermal boundary layers developed in the metal foam filled channel differ from those in an empty channel but may provide a rough idea of how the thermal boundary layers develop along the flow channel.

B. Test Setup

The 30 pores per inch (PPI) FeCrAl foam (Fig. 8.7a) is sandwiched between two 1 mm thick copper plates (used as the channel walls). A heating element is attached to the lower side of the channel wall, imposing constant heat flux, whilst the upper side is thermally insulated. The measurement is carried out under steady-state conditions. Fig. 8.7b shows a photograph of the test rig used in the present example.

The surface of the foam sample is sprayed with black paint. Thermal images of local solid temperature distribution are captured by an IR camera (AGEMA Thermovision 870). Since the side walls of the test rig were

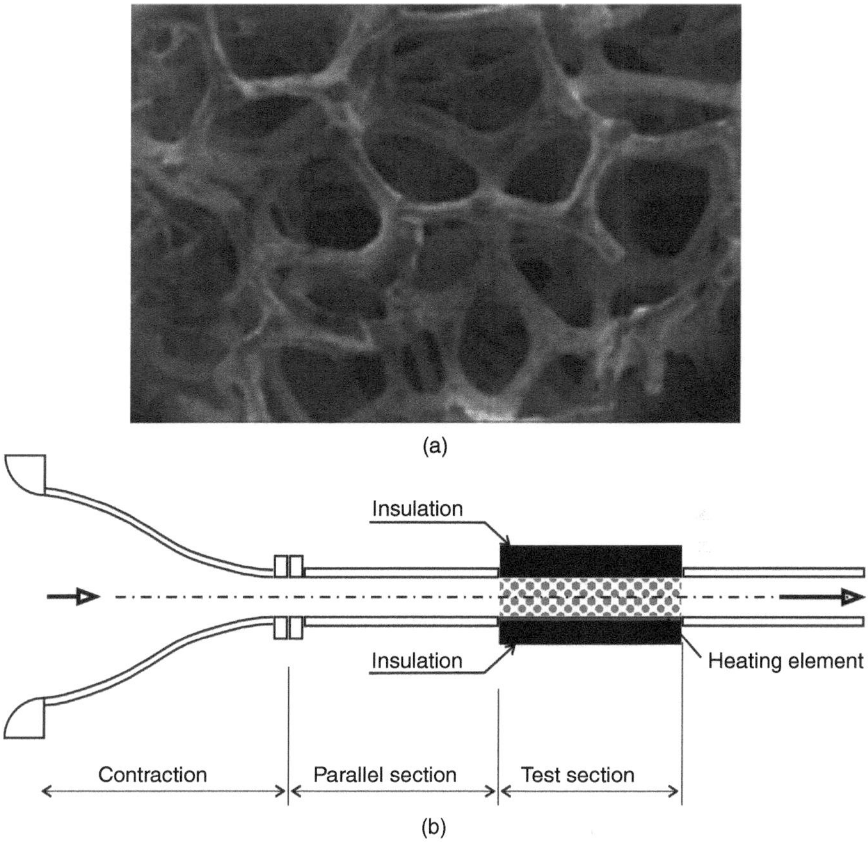

Figure 8.7 ***Test setup.*** (a) SEM image of open-celled metallic foam; (b) suction-type flow channel and test section.

originally made from Perspex, a special optical window is needed. For this purpose, a 3 mm thick quartz plate is used to replace the Perspex wall.

C. Discussion of Results

Fig. 8.8 shows the distribution of temperature on foam ligaments at two different Reynolds numbers based on the channel height (H). A brighter color indicates a higher temperature on foam ligaments which, with the assumption of thermal equilibrium, suggests a higher fluid flow temperature adjacent to the ligaments. Let δ_t denote the thickness of the thermal boundary layer.

For $\mathrm{Re}_H = 2500$, it is seen from Fig. 8.8a that the thermal boundary layer is clearly developed, with $\delta_t \sim \sqrt{x}$ where x is the axis along the direction of fluid flow and its origin coincides with the inlet of the test section. However, roughly at 3/4 length of the foam sample, its trend appears to

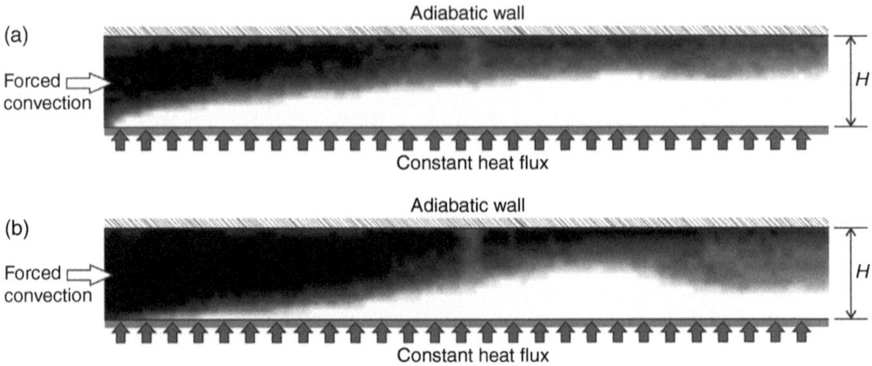

Figure 8.8 ***Thermal boundary layer in a metal foam filled channel for two different Reynolds numbers where H (= 12 mm) is channel height.*** (a) Low Reynolds number (Re_H = 2500); (b) high Reynolds number (Re_H = 10,000).

be disturbed due to the exit effect resulting from the sudden expansion of the mainstream flow into an empty channel and the discontinuity of heat imposed from the bottom of the channel.

At a higher Reynolds number, for example, Re_H = 10,000, the entry and exit effects become more pronounced. Since the foam length is relatively short compared to the channel height (or foam height), these effects dominate the entire thermal flow in the channel as clearly seen in Fig. 8.8b.

Example 8.2: Axial Fan Flow Impinging on a Flat Base Plate

A. Description of Problem

The high heat removal capability associated with an impinging jet has led to its widespread use in various engineering applications. Among these, using impinging jets for electronics cooling has been considered as one of the possible solutions for the increasing levels of heat flux emitted by electronic devices due to increasingly demanding compactness requirements (Saini and Webb, 2003; Chu et al., 2004; El-Sheikh and Garimella, 2000a,b; Brignoni and Garimella, 1999).

In conventional impinging jets, a steady high-momentum flow with the maximum velocity located at the center of the jet is issued through a circular nozzle/orifice, followed by impingement on a heat source. Extensive previous research on such impinging jets, including single and multiple arrays on flat/modified surfaces, can be found in the open literature (Saini and Webb, 2003; Chu et al., 2004; El-Sheikh and Garimella, 2000a,b; Brignoni and Garimella, 1999; Martin, 1977).

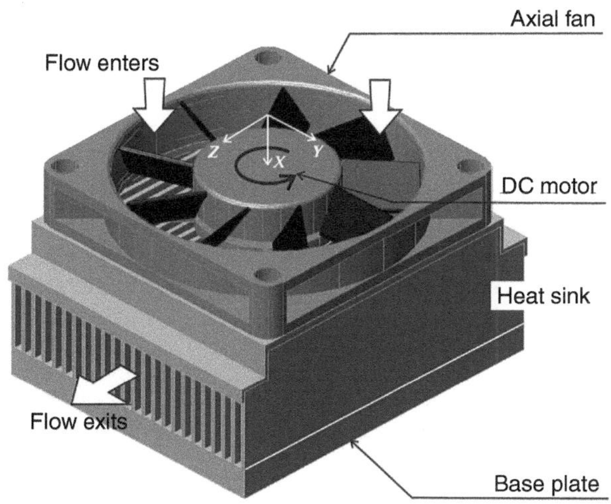

Figure 8.9 ***Typical heat sink unit for central processing unit (CPU) cooling.***

Although numerous efforts have been made to enhance the thermal performance using such impinging jets, cooling flows in conventional heat sinks are typically driven by an axial flow fan (Fig. 8.9). The mechanisms associated with the axial fan flow differ from those of the conventional impinging jet, because:

1. Due to the presence of a hub housing the DC driving motor, a large flow area at the central part of the impinging flow is displaced over roughly 25% of the total flow area and the impingement (direct impact of the emerged high-momentum flow) therefore occurs at target surface areas off from the axis of the driving fan;
2. The periodic impinging flow is highly unsteady, caused by rotating fan blades, compared to the mostly steady flow in a conventional jet;
3. The separation distance between the fan exit and the base plate (or fin height) is short, for example, typically half the fan diameter due to heat sink size restrictions;
4. No potential core exists, whilst its existence in a conventional jet is known to be influential on impinging heat transfer characteristics (especially for short separation distances).

With a flat plate placed downstream of the exit flow, the flow is decelerated as it approaches the plate, eventually stagnating on it. This implies changes of flow pattern within the fan exit-to-plate spacing (separation distance), resulting from the upstream penetration of the potential field. Furthermore, an increase in the intensity of the potential field penetration is expected as the separation distance is reduced. According to Huang and El-Genk (1994) and Baughn and Shimizu (1989), for conventional impinging jets, there exists an optimal separation distance for maximum stagnation heat transfer rate which is, in part, associated with the potential core of the jets. Conversely, no potential core is expected to exist in axial fan flows.

This case study aims to investigate experimentally the detailed patterns of axial fan flows in free exit and in impingement on a flat target plate. With the impingement, a particular emphasis is placed upon the "secondary stagnation region" on the target plate where the recirculated flows meet and interact, coinciding with the fan axis. The separation distance is systematically varied in the range of $0.2 \leq H/D \leq 2.0$, whilst the axial flow fan is rotated with 12.0 V, corresponding to a nominal Reynolds number (based on fan diameter D) of 9.0×10^3.

B. Test Setup

A schematic of the purpose-built test rig for examining the exit flow field associated with an axial flow fan is shown in Fig. 8.10. To facilitate the

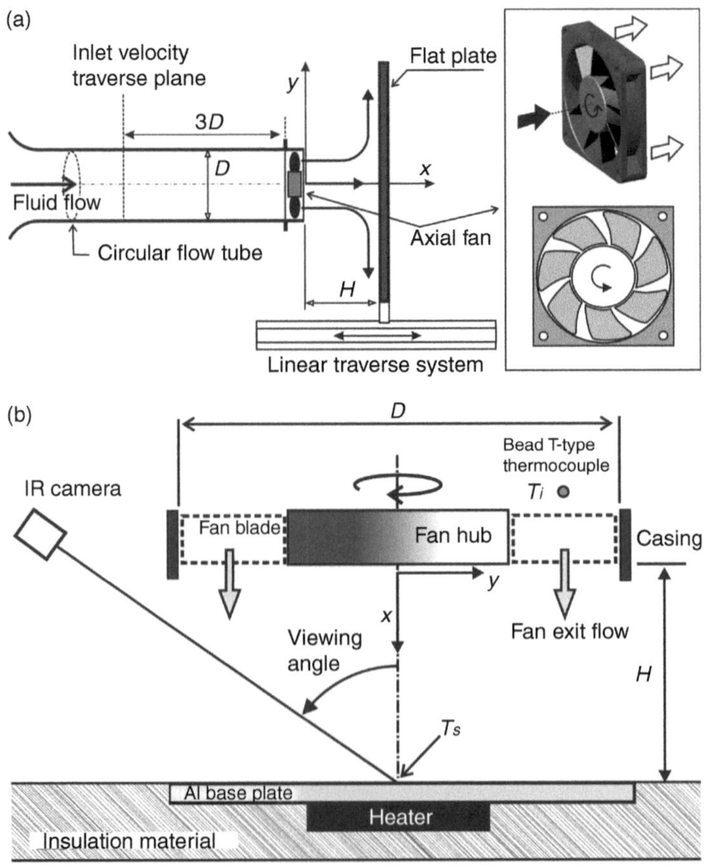

Figure 8.10 ***Schematic of test setup for an axial fan flow impinging on an Al block heated at the center.*** (a) Test rig; (b) IR setup.

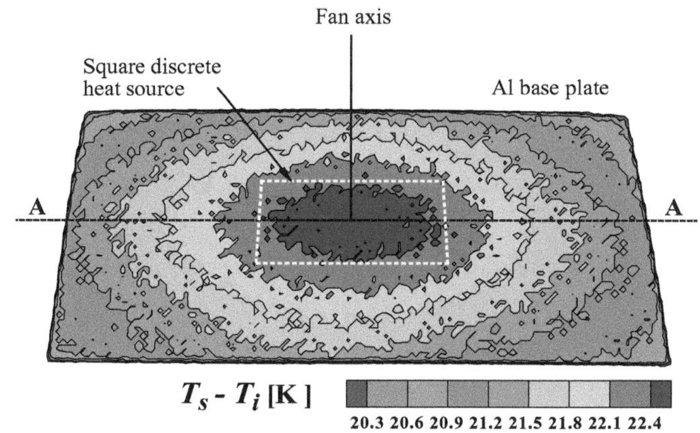

Figure 8.11 ***Temperature mapping on an Al base plate.***

measurement of mean flow velocity drawn by the axial fan, a circular flow tube is connected to the fan inlet. A Perspex flat plate with size of 0.3 m (height) × 0.3 m (width) × 0.01 m (thickness) is mounted on a linear traverse system, allowing systematic variation of the separation distance (H). The axial axis of the exit flow and the center of the plate coincide with the z-axis, as illustrated in Fig. 8.8.

The axial flow fan tested in this study has seven forward-swept fan blades and is ducted with a gap (tip clearance) between the blade tip and the casing. The ratio of tip clearance to blade span δ/S is 0.03. The fan steadily rotates, for example, with an error within ±0.5% at 2577 RPM for 12.0 V input from a DC power supply.

C. Discussion of Results

Fig. 8.11 shows the distribution of local temperature on the square Al block heated only at the center (as illustrated by the dotted square in the figure). It should be noted that the heat transfer results discussed here are taken as time averaged since the scan rate (or response time) of the IR camera, 10 Hz, is much slower than flow pattern changes caused by the fast-rotating axial fan (~ 40 Hz).

In general, along the A–A cross-section in Fig. 8.11, a higher temperature region is concentrated in the central part of the plate, followed by a monotonic decrease when moving away from the center. This high-temperature region is attributed to the presence of a hub housing the DC driving motor. A large flow area at the central part of the axial fan flow is displaced over roughly 25% of the total flow area and, therefore the impingement (direct impact of the emerged high-momentum flow) occurs at the Al block surface areas off from the driving fan axis.

8.5 CORRECTION FOR NONPERPENDICULAR VIEWING ANGLE

8.5.1 Background

As shown in Fig. 8.10b, for practical implementation of IR thermography, the viewing angle relative to the target surface plane may be nonperpendicular. In such cases, the image plane is no longer rectangular. In addition to the systematic reduction of the temperature values described in section 8.3.4, this section shows how to correct the distorted image due to the nonperpendicular viewing angle. One of the several possible methods is introduced.

With the assumption that a rectangular plate ($L \times W$) is focused by the IR camera at an arbitrary viewing angle (nonperpendicular) with respect to the flat target plane (Fig. 8.12a), the corresponding image taken (distorted due to the nonrectangular viewing angle) is illustrated in Fig. 8.12b. Here, the original plane is shown in the x–y coordinates and the corresponding image captured by the IR camera is shown in the ε–η coordinates.

A numerical code converts the physical plane in x–y coordinates into its image in the ε–η plane, if point-to-point correspondence using the following methodology is carried out.

8.5.2 Procedure

1. In total, eight points (a to h) on the rectangular original plane are selected, as illustrated in Fig. 8.13a where b, d, f, and h are the midpoints of the four sides, respectively. The corresponding points of a to h on the captured image are denoted as a' to h' as shown in Fig. 8.13b.

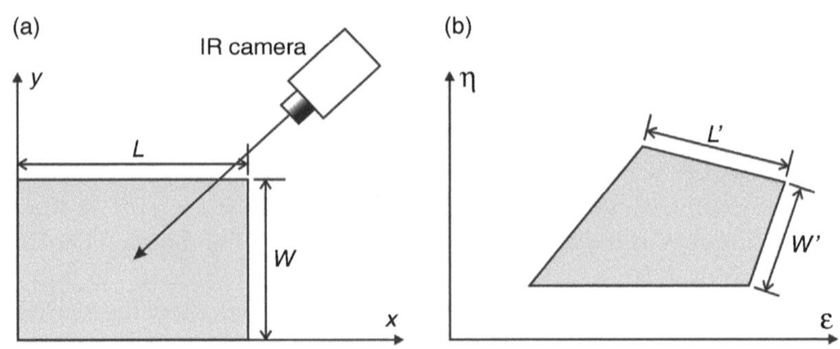

Figure 8.12 ***Image conversion with point-to-point matching.*** (a) Original image in x–y plane; (b) distorted image in ε–η plane.

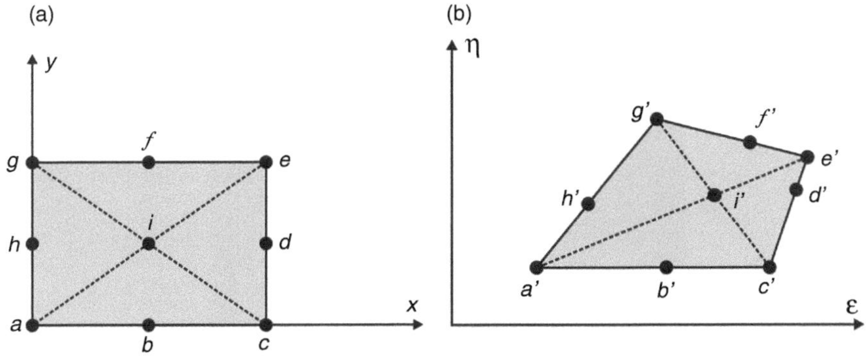

Figure 8.13 ***Image distortion due to nonperpendicular viewing angle.*** (a) Original plane; (b) distorted image.

2. By connecting *h'–d'* and *b'–f'*, the intersecting lines *h'd'* and *b'f'* must cross the center point *i'* as shown in Fig. 8.14a. The lines *h'–d'* and *b'–f'* divide the image into four small quadrilaterals, that is, *h'i'f'g'*, *i'd'e'f'*, *a'b'i'h'*, and *b'c'd'i'* (Fig. 8.14b). Repeating Step 1 for these four small quadrilaterals yields 12 more corresponding points.
3. By repeating Steps 1 and 2, the image can be divided into infinitely small quadrilaterals.

8.5.3 Example and Validation

Fig. 8.15a shows the captured raw thermal image on a rectangular plate (100 mm × 140 mm) subject to the heating of a hot air circular impinging jet. The image is taken at an arbitrary angle by an IR camera. After

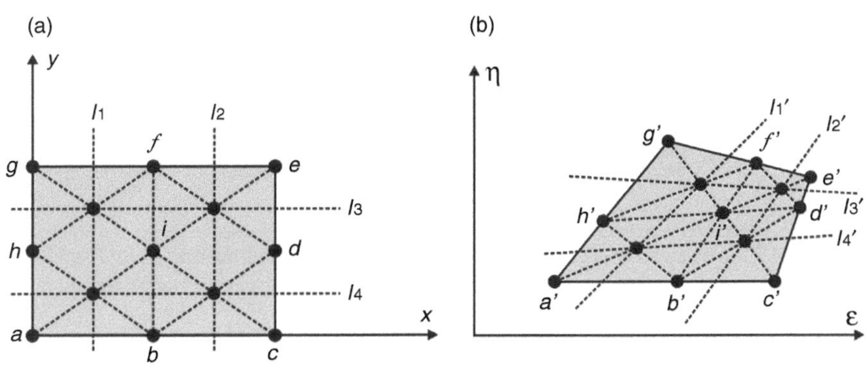

Figure 8.14 ***One-to-one point matching.*** (a) Original plane; (b) distorted image.

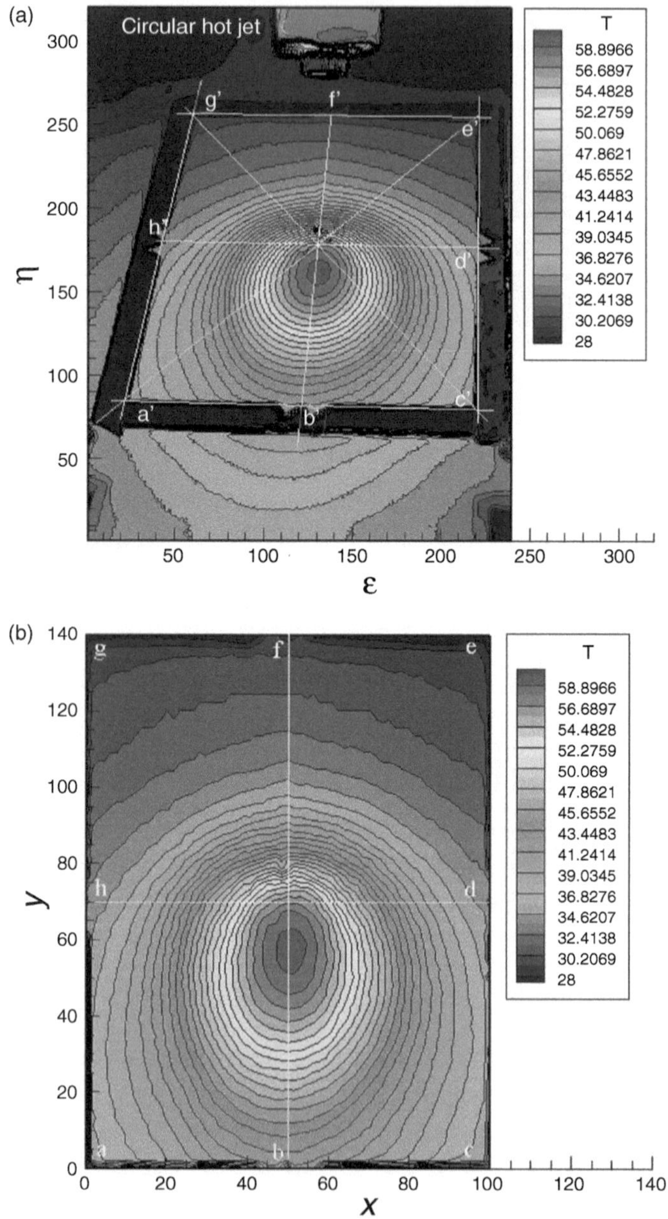

Figure 8.15 ***Correction of distorted IR thermal image with one-to-one point matching.*** (a) Original thermal image captured by IR camera; (b) corrected thermal image.

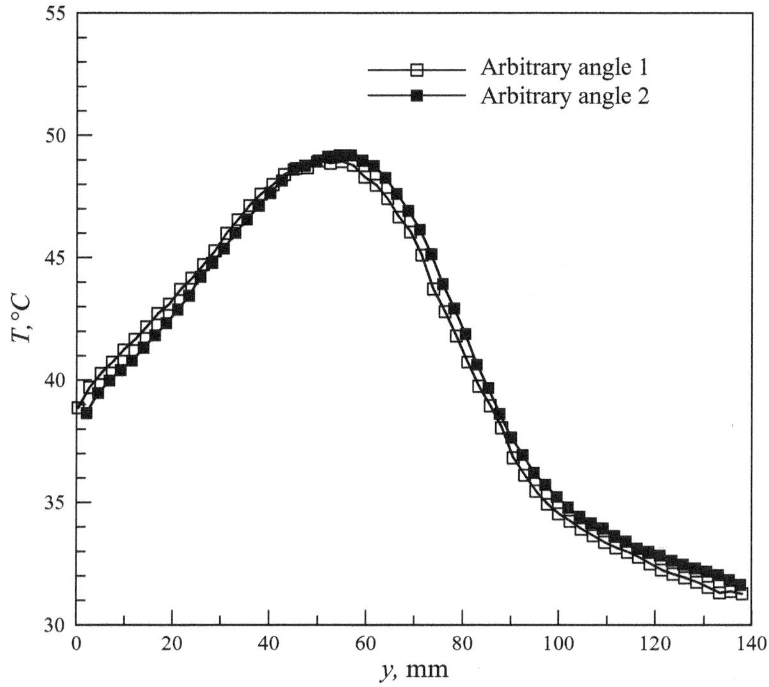

Figure 8.16 ***Comparison of corrected temperatures at two different viewing angles.***

correction using the method described above, the thermal image on the physical surface is converted to that shown in Fig. 8.15b.

According to the principle of the present correction method, the corrected image should be unique for images taken at different angles if the systematic errors caused by varying viewing angle are minimal. Fig. 8.16 presents the temperature profiles along the *y*-axis (Fig. 8.15b) at $x = 50$ mm for two different viewing angles. Excellent agreement between the two profiles is observed, validating thus the present algorithm (listed in section 8.5.4).

8.5.4 Fortran Code

An example of the viewing angle correction for Fig. 8.15, written by Fortran, is listed later.

```
MODULE START
     PARAMETER (NI=400, NJ=400)
     DIMENSION X(NI,NJ), Y(NI,NJ), XX(NI,NJ), YY(NI,NJ), XR(NI,NJ),
     & YR(NI,NJ), XT(NI,NJ),
     YT(NI,NJ), T(NI,NJ), TR(NI,NJ)
     COMMON X0, Y0, X1, Y1, X2, Y2, X3, Y3, X_x, Y_x
     INTEGER I, J, II, JJ, N, NT
     REAL X0, Y0, X1, Y1, X2, Y2, X3, Y3, X_x, Y_y, S0, S, L, M
     REAL K, C
END

PROGRAM MAIN
     USER START
     L=100
     M=140
     NT=6
C-------------------------------------------------------------------
        XX(1,1)=41.8868
        YY(1,1)=40.7713
        XX(2,1)=128.74
        YY(2,1)=42.1438
        XX(3,1)=214.298
        YY(3,1)=42.8301
        XX(1,2)=39.2941
        YY(1,2)=162.931
        XX(3,2)=212.353
        YY(3,2)=165.676
        XX(1,3)=36.7015
        YY(1,3)=285.09
        XX(2,3)=120.963
        YY(2,3)=288.522
        XX(3,3)=210.409
        YY(3,3)=291.267
        X0=XX(1,1)

        Y0=YY(1,1)

        X1=XX(3,1)

        Y1=YY(3,1)

        X2=XX(3,3)

        Y2=YY(3,3)

        X3=XX(1,3)

        Y3=YY(1,3)

        CALL INTSEC

        XX(2,2)=X_x

        YY(2,2)=Y_y
```

```
C----------------------------------------------------------------
      DO 100 N = 2, NT
C----------------------------------------------------------------
      DO J=2, 2**N, 2
      DO I=2, 2**N, 2
      X0=XX(I/2, J/2)
      Y0=YY(I/2, J/2)
      X1=XX(I/2+1, J/2)
      Y1=YY(I/2+1, J/2)
      X2=XX(I/2+1, (J+2)/2)!
      Y2=YY(I/2+1, (J+2)/2)
      X3=XX(I/2, (J+2)/2)!
      Y3=YY(I/2, (J+2)/2)
      CALL INTSEC
      X(I,J)=X_x
      Y(I,J)=Y_y
      END DO
      END DO
C----------------------------------------------------------------
      DO I=1, 2**N+1, 2
      DO J=1, 2**N+1, 2
      X(I, J)=XX((I+1)/2, (J+1)/2)
      Y(I, J)= YY((I+1)/2, (J+1)/2)
      END DO
      DO J=2, 2**N, 2
      X0=X(2, J)
      Y0=Y(2, J)
      X1=XX((I+1)/2, 1)
      Y1=YY((I+1)/2, 1)
      X2=X(2**N, J)
      Y2=Y(2**N, J)
      X3=XX((I+1)/2, 3)
      Y3=YY((I+1)/2, 3)
      CALL INTSEC
      X(I, J)=X_x
      Y(I, J)=Y_y
      END DO
      END DO
C----------------------------------------------------------------
      DO I=2, 2**N, 2
      DO J=1, 2**N+1, 2
      X0=X(I, 2)
      Y0=y(I, 2)
      X1=XX(1, (J+1)/2)
      Y1=YY(1, (J+1)/2)
      X2=X(I, 2**N)
      Y2=Y(I, 2**N)
      X3=XX(3, (J+1)/2)
      Y3=YY(3, (J+1)/2)
```

```
      CALL INTSEC
      X(I, J)=X_x
      Y(I, J)=Y_y
      END DO
      END DO
C-----------------------------------------------------------------
      DO J=1, 2**N+1
      DO I=1, 2**N+1
      XX(I, J)=X(I, J)
      YY(I, J)= Y(I, J)
      END DO
      END DO
100 CONTINUE
C-----------------------------------------------------------------
      OPEN(1, FILE=IR.DAT)
      DO 105 I=1, 240
      DO 105 J=1, 320
      READ(1, *) XR(I, J), YR(I, J), TR(I, J)
105 CONTINUE
      CLOSE(1)
C-----------------------------------------------------------------
      DO 200 J=1, 2**NT+1
      DO 200 I=1, 2**NT+1
      SO=100
      DO JJ=1, 320
      DO II=1, 240
      S=SQRT((X(I, J)-XR(II, JJ))* *2+(Y(I, J)-YR(II, JJ))**2)
      IF(S, LE, SO) THEN
      T(I, J)=TR(II, JJ)
      SO=S
      END IF
      END DO
      END DO
200 CONTINUE
C-----------------------------------------------------------------
      DO I=1, 2**NT+1
      YT(I, 1)=0
      XT(I, 1)=L/(2**NT)*(I-1)
      END DO
      DO J=1, 2**NT+1
      XT(1, J)=0
      YT(1, J)=M/(2**NT)*(J-I)
      END DO
      DO 300 J=2, 2**NT+1
      DO 300 I=2, 2**NT+1
      XT(I, J)=L/(2**NT)*(I-1)
      YT(I, J)=M/(2**NT)*(J-1)
300 CONTINUE
C-----------------------------------------------------------------
```

```
400 FORMAT(1P5E15. 6)
      OPEN(2, FILE=RECTANGULAR. DAT)
      WRITE(2, *) 'VARIABLES="X" "Y" "T"'
      WRITE(2, *) 'ZONE I=', 2 * *NT+1, 'J=', 2* *NT+1
      DO J=1, 2**NT+1
      DO I=1, 2**NT+1
      WRITE(2, 400) XT(I, J), YT(I, J), T(I, J)
      END DO
      END DO
      CLOSE(2)
C-------------------------------------------------------------------
      OPEN(3, FILE= 'ORIGIN. DAT')
      WRITE(3, *) 'VARIABLES="X" "Y" "T"'
      WRITE(3, *) 'ZONE I=', 2**NT+1, 'J=',2**NT+1
      DO J=1, 2* *NT+1
      DO I=1, 2* *NT+1
      WRITE(3, 400) X(I, J), Y(I, J), T(I, J)
      END DO
      END DO
      CLOSE(3)
END

SUBROUTINE INTSEC
     USE START
     K=(Y3-Y1)/(X3-X1)-(Y2-Y0)/(X2-X0)
     C=(X2*Y0-X0*Y2)/(X2-X0)-(X3*Y1-X1*Y3)/(X3-X1)
     X_x=C/K
     Y_y=(Y2-Y0)/(X2-X0)*X_x+(X2*Y0-X0*Y2)/(X2-X0)
END
```

Example 8.3: Slot Jet Impingement on a Concave Surface

A. Background

Impinging jet cooling (or heating) has long been studied due to its superior heat transfer capability to other convective heat transfer schemes. Either 2D (slot) jet or 3D (round) jet impinging on a flat plate serves as a fundamental configuration amongst others. Consequently, a multitude of studies have been devoted to this configuration (Martin, 1977; Jambunathan et al., 1992; Gauntner et al., 1970; Gardon and Akfirat, 1965; Reungoat et al., 2007).

When approaching the flat plate, the jet is decelerated and eventually stagnates on the flat plate, followed by radial deflection. After stagnation, the jet is radially (or laterally) deflected and undergoes contraction and acceleration, followed by a deceleration. A laminar boundary layer develops from the stagnation point and its thickness grows. It should be noted that regardless of the jet conditions such as either a turbulent or laminar jet approaching the flat plate, a laminar boundary layer starts to develop. The wall parallel velocity component, initially increasing, reaches a maximum value at a certain distance from the stagnation point. The stabilizing effect of acceleration keeps the boundary layer laminar. After which, transition to turbulence generally will occur in the deceleration region (Martin, 1977).

In high heat flux applications such as internal cooling of gas turbine blades, jet impingement at the leading edge of the blades is on a concave surface. It has been reported that the variations of local heat transfer including at the stagnation point are similar to those observable on a flat plate (Gulati et al., 2009; Yang et al., 1999; Gau and Chung, 1991). Nevertheless, higher average heat transfer has been obtained from a concave surface than from a flat plate. Centrifugal instabilities, developed as a result of the curvature of the concave surface, induce the formation of so-called Görtler vortices, leading to the enhanced local heat transfer (Yang et al., 1999; Lee et al., 1999; Gau and Chung, 1991).

Local temperature (or heat transfer) mapping on a flat plate subjected to either a slot jet or round jet using an IR camera is relatively straightforward whilst still requiring a series of calibrations for viewing angles (eg, α and β) as illustrated in Fig. 8.17, which shows a test setup for a round impinging jet on a flat plate. Since there is an optical obstruction by the test rig, for example, a round jet nozzle, the IR camera needs to be positioned at the angles in Fig. 8.17, which can cause errors. The two viewing angles are coupled but it is still possible to lump all factors that affect temperature values together through calibration regardless of the target surface geometries, for example, flat or curved. This example demonstrates how the surface geometry can cause errors but these can be overcome by performing a series of calibrations.

B. Experimental Details

Fig. 8.18 shows a schematic of the test rig consisting of a slot jet impinging on a flat plate (0.5 m × 0.5 m). The flat plate was fabricated from a 10 mm thick acrylic material ($k \sim 0.2$ W/mK). For heat transfer measurement, a circular film heating element ($D = 0.3$ m with thickness of 0.27 mm) was attached to the target surface, emitting a constant heat flux (Fig. 8.18). Two foil and one bead-type precalibrated T-type thermocouples were used to measure the temperature at the stagnation point on the flat plate, on the back of the target surface (to estimate heat loss), and at the round jet exit, respectively.

Air at ambient conditions drawn by a centrifugal fan controlled via an inverter was discharged from the slot with a fixed width of $B = 0.026$ m. The target surface was mounted on a linear traverse system so that the jet exit-to-target surface distance, H, along the jet axis could be varied systematically.

The jet Reynolds number based on the slot jet width (B) and jet centerline exit velocity (w_e) is defined as:

$$\mathrm{Re}_{Dj} = \frac{\rho w_e B}{\mu} \tag{8.13}$$

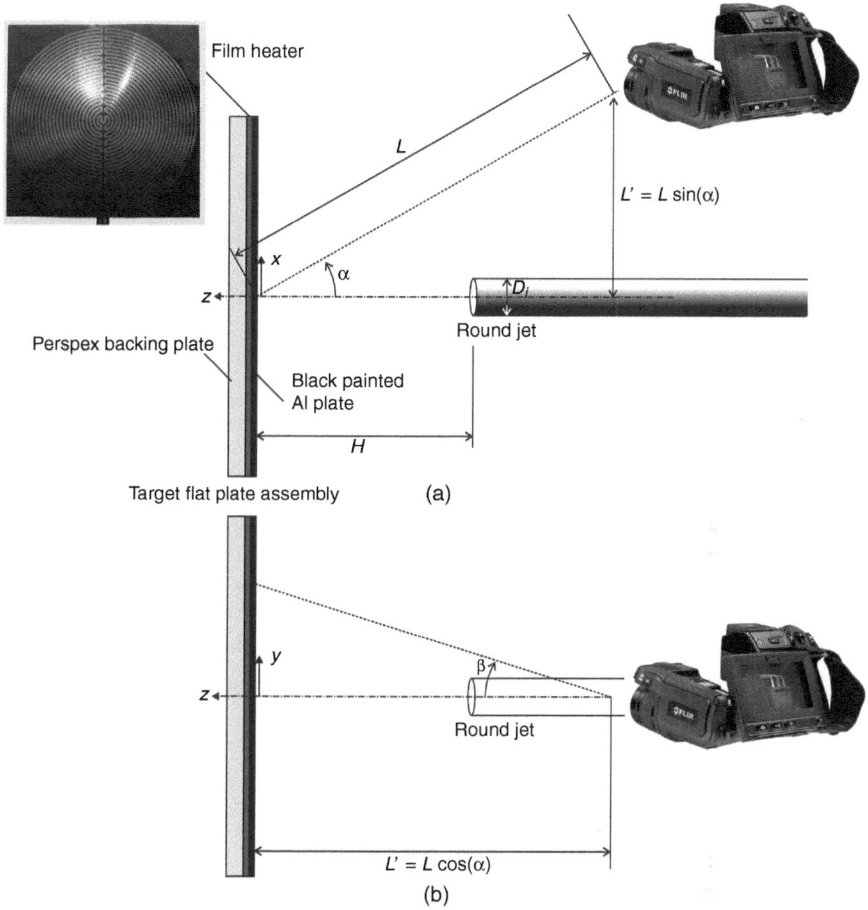

Figure 8.17 ***Schematic of the test setup for the round jet impingement cooling on a uniformly heated flat plate and an IR camera.*** (a) In the *z*–*x* plane; (b) in the *y*–*z* plane.

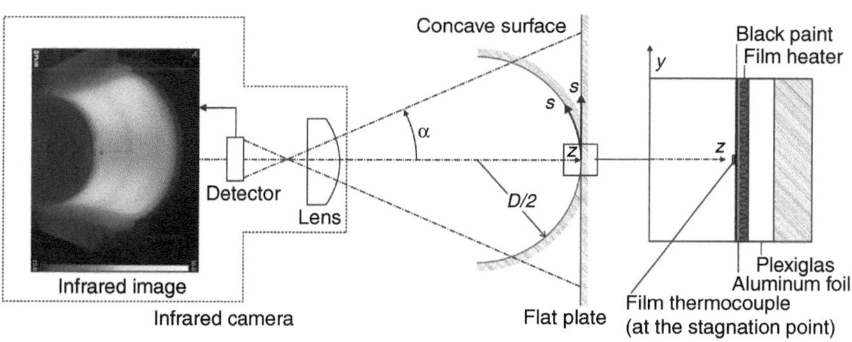

Figure 8.18 ***Schematic of IR temperature measurement on a flat plate and concave surface.***

During the entire experiment, Re_{Dj} was fixed at 20,000. The heat transfer characteristics were evaluated using the Nusselt number defined based on the round jet diameter D_j as:

$$\mathrm{Nu}_{Dj} = \frac{h_s}{k_f / B} \tag{8.14}$$

where

$$h_s = \frac{q''_{net}}{T_s - T_j} \tag{8.15}$$

Here, q''_{net} is the net heat flux, k_f is the thermal conductivity of air, whilst T_s and T_j are the local surface temperature (or T_{IR}) and jet temperature at the jet exit, $z = 0$, respectively. The net heat flux, q''_{net}, is obtained through energy balance:

$$q''_{net} = q''_t - q''_{cond} - q''_{rad} \tag{8.16}$$

where q''_t is the total heat flux emitted from the film heating element, and q''_{cond} and q''_{rad} are the conduction and the radiation heat loss, respectively. The conduction heat loss through the backside of the target surface was estimated using Fourier's law and found to be 3.5%. The radiation loss is calculated using the Stefan–Boltzmann law:

$$q''_{rad} = \varepsilon\sigma\left(T_s^4 - T_a^4\right) \tag{8.17}$$

Here, ε is the emissivity, σ is the Stefan–Boltzmann constant, and T_a is the ambient temperature. The emissivity was taken to be 0.94 as estimated by an IR camera (FLIR T-Series).

The resolution of the temperature readings from the temperature scanner was ±0.1 K for each thermocouple. The errors associated with the multimeter readings for the input voltage to, and the electrical resistance of, the film heater were estimated to be ±0.09 V and ±0.27 Ω, respectively, resulting in a total heat flux error of ±11.1 W/m^2. The uncertainty of the net heat flux was estimated to be within 1.1%. Therefore, the uncertainty of the Nusselt number was calculated to be within 3.3%.

Despite the benefits of IR thermography, it also has limitations in the form of many parameters which can cause measurement error; this includes parameters to consider during calibration (object emissivity, reflective temperature, ambient temperature, relative humidity, and distance). For the current study, an IR camera (FLIR T-Series) was used to measure the local heat transfer distribution on a target surface subjected to an impinging round jet. The emissivity was calibrated using the contact method described in Usamentiaga et al. (2014) as follows:

- The target surface was heated without a jet, until it reached temperatures similar to that during jet impingement.

- A film thermocouple flush-mounted on the target surface (at the stagnation point; Fig. 8.16) was used to obtain a reference temperature.
- Once the target surface had reached temperatures similar to that during jet impingement, a reference temperature was measured with the thermocouple, then the IR camera emissivity was changed until it detected a temperature matching with the reference temperature at the same location.

A second IR camera emissivity calibration method called the reference emissivity material method is described in Usamentiaga et al. (2014) and can also be used.

The reflected temperature was estimated using a method also described in Usamentiaga et al. (2014) and ASTM E1862-97(2002)e1 (1997) as follows:

- A crumpled and re-flattened piece of aluminum foil was used as a reflector.
- The re-flattened aluminum foil was positioned in the field of view of the IR camera.
- The temperature of the foil was measured with the emissivity and distance set at 1 and 0 on the camera.
- The reflected temperature was measured by applying the temperature of the foil to the surface; the resulting temperature value is the reflected temperature.

Amongst other parameters causing a measurement error while using an IR camera, the viewing angle is commonly encountered, due to the fact that it is not always possible to align the camera with the surface to be measured. To account for the effect of the viewing angle (only one viewing angle is involved for the slot jet impingement), the temperature distribution on the target surface was captured with the IR camera for the isothermal case (without an impinging jet). IR images were captured while constant uniform heat flux (36.79 W/m^2, 229.94 W/m^2, and 331.12 W/m^2) was maintained on the target surface. The IR images were processed using FLIR Tools+ software in order to extract the temperature distribution along the *s*-axis, and compared to the thermocouple (at the stagnation point) measured temperature. The temperature of the thermocouple at 0° relative viewing angle was matched to the thermocouple temperature at the same location by adjusting the emissivity setting on the IR camera.

Fig. 8.19a, b show that when the camera (lens) is positioned 1.0 m away from the flat plate (which is the actual distance set up), deviation of the temperature values obtained from the IR camera from those measured by the thermocouple occurs at the viewing angle of 4.45° for the flat plate (defined to be $T_{IR}/T_{TC} \geq 0.99$), regardless of the applied heat flux values. After which, the IR camera sees the local temperature being lower than the actual temperature value. To correct errors associated with the viewing angle, the curves in Fig. 8.19 were used to derive a fourth-order polynomial expression of the curve trend, which was used to correct the deviation.

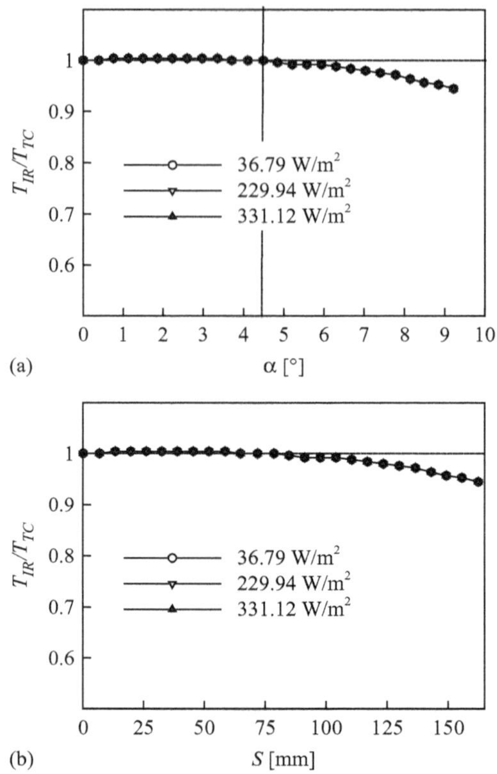

Figure 8.19 ***Effect of viewing angle on IR temperature normalized by thermocouple temperature (T_{IR}/T_{TC}) on a flat plate versus: (a) the IR camera relative viewing angle (α); (b) the radial location (s).***

Fig. 8.20 shows how the viewing angle to the concave surface affects the temperature readings from the IR camera. In comparison to the flat plate case, the valid viewing angle, that is, $T_{IR}/T_{TC} \geq 0.99$, is shortened to be less than 3.5° (cf, 4.5° on the flat plate). After which, the IR camera sees the local temperature being lower than the actual temperature value whose extent is worse than that on the flat plate. The concave surface configures further oblique viewing angles to the IR camera.

C. Discussion of Results

Higher average heat transfer can be achieved from a concave surface than from a flat plate, due to centrifugal instabilities on a concave surface resulting in the so-called "Görtler vortices" (Yang et al., 1999; Gau and Chung, 1991; Lee et al., 1999).

Fig. 8.21a–c compare the local heat transfer distribution on the flat plate ($D/B = \infty$) with that on the small concave surface ($D/B = 9.0$) at three selected H/B values, for example, $H/B = 4.0$, 6.0, and 12.0, obtained

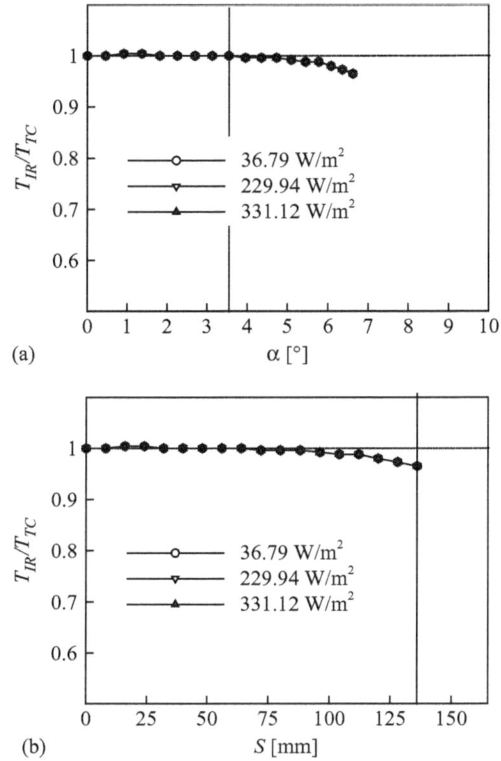

Figure 8.20 ***Effect of viewing angle on IR temperature normalized by thermocouple temperature (T_{IR}/T_{TC}) on a concave surface versus: (a) the IR camera relative viewing angle (α); (b) the radial location (s).***

using the precalibrated IR camera. In all cases, the highest heat transfer is located at the stagnation point and decreased along the *s*-axis (Fig. 8.21). At $H/B = 4.0$, which is inside the potential core of the present slot jet, there exists a second peak in heat transfer on the flat plate as a result of the transition from laminar to turbulent flow (Gardon and Akfirat, 1965; Zhou and Lee, 2007; Katti and Prabhu, 2008; Nirmalkumar et al., 2011; Gulati et al., 2009; Wang et al., 2014; Gau and Chung, 1991). For the same reason, there might be a second peak on the concave surface, which, however, cannot be fully observed due to the viewing angle distortion near the edge of the concave surface. At $H/B = 12.0$, the local heat transfer distribution on the concave surface is almost constant whereas that on the flat plate monotonically decreases from the stagnation point along the *s*-axis. It appears that the area-average heat transfer over the flat plate is lower than that over the concave surface, which is consistent with previous results (Yang et al., 1999; Gau and Chung, 1991; Lee et al., 1999).

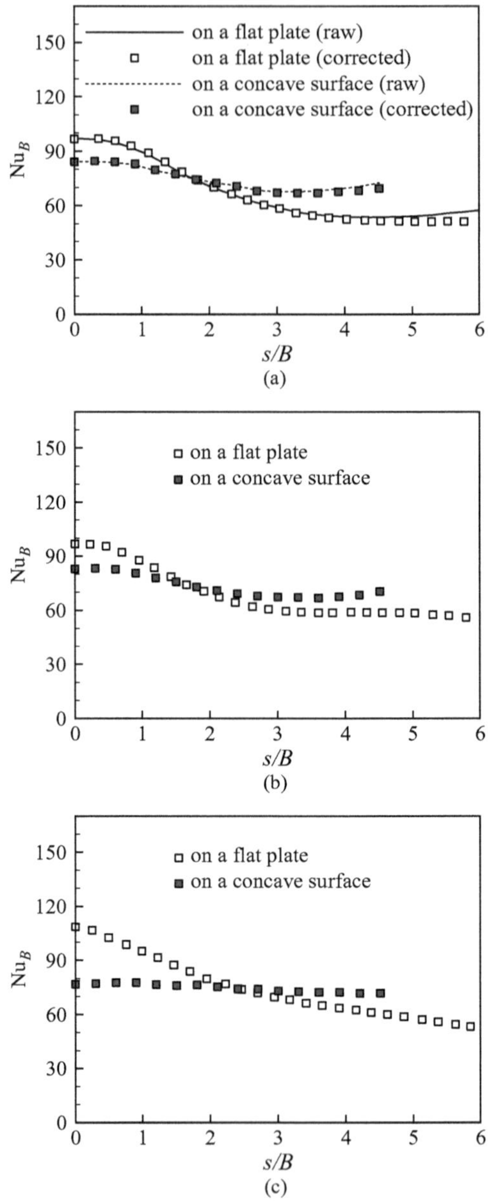

Figure 8.21 ***Local heat transfer along the flat plate and concave surface*** **(D/B = *9.0)*.** (a) $H/B = 4.0$; (b) $H/B = 6.0$; (c) $H/B = 12.0$.

REFERENCES

ASTM E1862-97(2002)e1, 1997. Standard Test Methods for Measuring and Compensating for Reflected Temperature Using Infrared Imaging Radiometers, ASTM International, West Conshohocken, PA, USA.

Baughn, J.W., Shimizu, S., 1989. Heat transfer measurements from a surface with uniform heat flux and an impinging jet. J. Heat Transf. 111, 1096–1098.

Beck, J., Blackwell, B., Clair, C., 1985. Inverse Heat Conduction: Ill-Posed Problems. Wiley-Interscience, Hoboken, NJ.

Brignoni, L.A., Garimella, S.V., 1999. Experimental optimization of confined air jet impingement on a pin fin heat sink. IEEE T. Compon. Pack. T. 22, 399–404.

Carlomagno, G.M., Cardone, G., 2010. Infrared thermography for convective heat transfer measurements. Exp. Fluids 49 (6), 1187–1218.

Cheng, H.K., Chang, S.M., 1990. Application of the hybrid method to inverse heat conduction problems. Int. J. Heat Mass Tran. 33, 621–628.

Childs, P.R.N., 2001. Practical Temperature Measurements. Butterworth-Heinemann, Oxford.

Chu, R.C., Simons, R.E., Ellsworth, M.J., Schmidt, R.R., Cozzolino, V., 2004. Review of cooling technologies for computer products. IEEE T. Device Mat. Re. 4, 568–585.

Cole, K.D., Beck, J.V., Hiji-Sheikh, A., Litkouhi, B., 2010. Heat Conduction Using Green's Functions, Series in Computational and Physical Processes in Mechanics and Thermal Sciences. Taylor & Francis Group, London.

El-Sheikh, H.A., Garimella, S.V., 2000a. Enhancement of air jet impingement heat transfer using pin-fin heat sink. IEEE T. Compon. Pack. T. 23, 300–308.

El-Sheikh, H.A., Garimella, S.V., 2000b. Heat transfer from pin-fin heat sinks under multiple impinging jets. IEEE T. Adv. Packaging 23, 113–120.

Fernandes, A.P., dos Sandos, M.B., Guimaraes, G., 2015. An analytical transfer function method to solve heat conduction problems. Appl. Math. Model. 36, 6897–6914.

Frank, I., 1963. An application of least squares method to the solution of the inverse problem of heat conduction. J. Heat Transf. 85C, 378–379.

Gardon, R., Akfirat, J.C., 1965. The role of turbulence in determining the heat-transfer characteristics of impinging jets. Int. J. Heat Mass Tran. 8 (10), 1261–1272.

Gau, C., Chung, C.M., 1991. Surface curvature effect on slot-air jet impingement cooling flow and heat transfer process. J. Heat Transf. 113 (4), 858–864.

Gauntner, J., Livingood, N.B., Hrycak, P., 1970. Survey of literature on flow characteristics of a single turbulent jet impinging on a flat plate. NASA TN D-5652 Lewis Research Center, USA. NASA TN D-5652.

Gaussorgues, G., 1994. Infrared Thermography. Springer, Dordrecht, The Netherlands.

Gruner, K.-D., 2003. Principles of Non-Contact Temperature Measurement. Raytek, Santa Cruz, CA.

Gulati, P., Katti, V., Prabhu, S.V., 2009. Influence of the shape of the nozzle on local heat transfer distribution between smooth flat surface and impinging air jet. Int. J. Therm. Sci. 48 (3), 602–617.

Huang, L., El-Genk, M.S., 1994. Heat transfer of an impinging jet on a flat surface. Int. J. Heat Mass Tran. 37, 1915–1923.

Jambunathan, K., Lai, E., Moss, M.A., Button, B.L., 1992. A review of heat transfer data for single circular jet impingement. Int. J. Heat Fluid Flow 13 (2), 106–115.

Jewett, J.W., Serway, R.A., 2010. Physics for Scientists and Engineers with Modern Physics. Brooks Cole, Pacific Grove, CA.

Katti, V., Prabhu, S.V., 2008. Experimental study and theoretical analysis of local heat transfer distribution between smooth flat surface and impinging air jet from a circular straight pipe nozzle. Int. J. Heat Mass Tran. 51 (17–18), 4480–4495.

Kumar, R., Jayesh, P., Sahoo, N., 2012. Analysis of one dimensional inverse heat conduction problem: a review. Int. J. Mechanical Industrial Engineering 2, 15–20.

Lee, D.H., Chung, Y.S., Won, S.Y., 1999. Technical note: the effect of concave surface curvature on heat transfer from a fully developed round impinging jet. Int. J. Heat Mass Tran. 42 (13), 2489–2497.

Martin, H., 1977. Heat and mass transfer between impinging gas jets and solid surfaces. Advances in Heat Transfer 13, 1–60.

Nirmalkumar, M., Katti, V., Prabhu, S.V., 2011. Local heat transfer distribution on a smooth flat plate impinged by a slot jet. Int. J. Heat Mass Tran. 54 (1–3), 727–738.

Optotherm Thermal Imaging, Emissivity in the Infrared. Available from: <http://optotherm.com/emiss-effects.htm>.

Reungoat, D., Riviere, N., Faure, J.P., 2007. 3D PIV and PLIF measurement in turbulent mixing round jet impingement. J. Vis. 10 (1), 99–110.

Saini, M., Webb, R., 2003. Heat rejection limits of air cooled plane fin heat sinks for computer cooling. IEEE T. Compon. Pack. T. 26, 71–79.

Smits, A.J., Lim, T.T., 2003. Flow Visualization: Techniques and Examples. Imperial College Press, London.

Stolz, G., 1960. Numerical solution to inverse problem of heat conduction for simple shapes. J. Heat Transf. 82C, 20–26.

Usamentiaga, R., Venegas, P., Guerediaga, J., Vega, L., Molleda, J., Bulnes, F.G., 2014. Infrared thermography for temperature measurement and non-destructive testing. Sensors – Basel 14 (7), 12305–12348.

Vollmer, M., Mollmann, K.P., 2010. Infrared Thermal Imaging: Fundamentals, Research and Applications. Wiley, Hoboken, NJ.

Wang, X.L., Motala, D., Lu, T.J., Song, S.J., Kim, T., 2014. Heat transfer of a circular impinging jet on a circular cylinder in crossflow. Int. J. Therm. Sci. 78, 1–8.

Yang, G., Choi, M., Lee, J.S., 1999. An experimental study of slot jet impingement cooling on concave surface: effects of nozzle configuration and curvature. Int. J. Heat Mass Tran. 42 (12), 2199–2209.

Zhou, D.W., Lee, S.-J., 2007. Forced convective heat transfer with impinging rectangular jets. Int. J. Heat Mass Tran. 50 (9–10), 1916–1926.

INDEX

J

K

L

Q

R

U

V

W

Y

Z

Kolibri 360 GmbH
Ferdinand-Jühlke-Straße 7
99095 Erfurt, Deutschland
produktsicherheit@kolibri360.de